Society for the Study of Human Biology Series

THE CHANGING FACE OF DISEASE:

IMPLICATIONS FOR SOCIETY

Published Symposia of the Society for the Study of Human Biology

10 Biological Aspects of Demography

Edited by W. Brass

11 Human Evolution

Edited by M. H. Day

12 Genetic Variation in Britain

Edited by D. F. Roberts and E. Sunderland

13 Human Variation and Natural Selection

Edited by D. Roberts (Penrose Memorial Volume reprint)

14 Chromosome Variation in Human Evolution

Edited by A. J. Boyce

15 Biology of Human Foetal Growth

Edited by D. F. Roberts

16 Human Ecology in the Tropics

Edited by J. P. Garlick and R. W. J. Keay

17 Physiological Variation and its Genetic Base

Edited by J. S. Weiner

18 Human Behaviour and Adaption

Edited by N. J. Blurton Jones and V. Reynolds

19 Demographic Patterns in Developed Societies

Edited by R. W. Horns

20 Disease and Urbanisation

Edited by E. J. Clegg and J. P. Garlick

21 Aspects of Human Evolution

Edited by C. B. Stringer

22 Energy and Effort

Edited by G. A. Harrison

23 Migration and Mobility

Edited by A. J. Boyce

Numbers 1–9 were published by Pergamon Press, Headington Hill Hall, Headington, Oxford OX3 0BY. Numbers 10–24 were published by Taylor & Francis Ltd, 10–14 Macklin Street, London WC2B 5NF. Numbers 25–40 were published by Cambridge University Press, The Pitt Building, Trumpington Street Cambridge CB2 1RP. Further details and prices of back-list numbers are available from the Secretary of the Society for the Study of Human Biology.

Society for the Study of Human Biology Series

THE CHANGING FACE OF DISEASE:

IMPLICATIONS FOR SOCIETY

EDITED BY

Nick Mascie-Taylor
Jean Peters
Stephen T. McGarvey

CRC PRESS

Boca Raton   London   New York   Washington, D.C.

Library of Congress Cataloging-in-Publication Data

Catalog record is available from the Library of Congress

Visit the CRC Press Web site at www.crcpress.com

© 2004 by CRC Press LLC

No claim to original U.S. Government works
International Standard Book Number 0-415-32280-4
Printed in the United States of America 1 2 3 4 5 6 7 8 9 0
Printed on acid-free paper

Contents

Figures

Tables

Contributors

Gemiliano D. Aligui
Philippine Council for Health Research and Development Bicutan
Tagig
MetroManila
The Philippines

George J. Armelagos
Department of Anthropology
Emory University
Atlanta GA 30309
USA

Emily S. Barrett
Department of Anthropology
Harvard University
Cambridge MA 02138
USA

John Blangero
Department of Genetics
Southwest Foundation for Biomedical Research
PO Box 760549
San Antonio TX 78245-0549
USA

Hélène Carabin
Department of Biostatistics and Epidemiology
Oklahoma University Health Sciences Center
Oklahoma City OK 73104
USA

Peter T. Ellison
Department of Anthropology
Harvard University
Cambridge MA 02138
USA

Brian J. Ford
Rothay House
6 Mayfield Road
Eastrea
Cambridgeshire PE7 2AY
UK

Robin Goodwin
Department of Human Sciences
Brunel University
Uxbridge
Middlesex
UK

Elaine A. Hills
Department of Anthropology
University at Albany
State University of New York
Albany NY 12222
USA

Stephen J. Kunitz
Department of Community and Preventive Medicine
Box 644
University of Rochester Medical Center
601 Elmwood Avenue
Rochester NY 14642
USA

Jonathan D. Kurtis
Brown University
International Health Institute
Providence RI 02912
USA

S. W. Lindsay
School of Biological and Biomedical Sciences
University of Durham
Science Laboratories
South Road
Durham DH1 3LE
UK

Stephen T. McGarvey
International Health Institute
Box G-B497
Brown University
Providence RI 02912
USA

R. McGready
Shoklo Malaria Research Unit
P.O. Box 46
Mae Sot 63110
Thailand;
Faculty of Tropical Medicine
Mahidol University
Bangkok
Thailand;
Centre for Tropical Medicine
Nuffield Department of Medicine
John Radcliffe Hospital
Headington
Oxford
UK

Remigio Olveda
Research Institute for Tropical Medicine
Department of Health
Muntinlupa
MetroManila
The Philippines

Jean Peters
Section of Public Health, University of Sheffield
School of Health and Related Research (ScHARR)
Regent Court
30 Regent Street
Sheffield S1 4DA

Lisa Sattenspiel
Department of Anthropology
107 Swallow Hall
University of Missouri
Columbia MO 65211
USA

Lawrence M. Schell
Departments of Anthropology and Epidemiology
University at Albany
State University of New York
Albany NY 12222
USA

John L. VandeBerg
Office of the Director
Southwest National Primate Research Centre
7620 NW Loop 410
San Antonio TX 78227
USA;
Department of Genetics
Southwest Foundation for Biomedical Research
PO Box 760549
San Antonio TX 78245-0549
USA

G. E. L. Walraven
Medical Research Council Laboratories
P.O. Box 273
Banjul
The Gambia

Nicholas J. Wareham
University of Cambridge Department of Public Health and Primary Care
Institute of Public Health
Robinson Way
Cambridge CB2 2SR
UK

Sarah Williams-Blangero
Department of Genetics
Southwest Foundation for Biomedical Research
P.O. Box 760549
San Antonio TX 78245–0549
USA

Arve Lee Willingham III
WHO/FAO Collaborating Center for Parasitic Zoonoses
Danish Centre for Experimental Parasitology
Royal Veterinary and Agricultural University
Frederiksberg
Denmark

Introduction

It is commonplace now to criticize the hopeful assertions of 30 years ago that infectious diseases were on the wane. These incorrect prognostications were based on a bias towards the developed world as well as a lack of understanding of the extant information on the social epidemiology of infectious diseases in North America and northwestern Europe. One of the key contributions of this book is the description and analysis of how these incorrect assertions were and remain incorrect in the contemporary world. The chapters embrace a more balanced human ecological perspective to diseases that includes many levels of analysis required for accurate recording and understanding of the current status of human health and disease.

This book, the result of the Society for the Study of Human Biology's 43rd Symposium, reflects mostly the human population biology, biological anthropology and biomedical and public health perspectives and studies of many of the authors. But the themes and perspectives in the book will also prepare the reader for the increasingly more standardized and influential area of scholarship known as the global burden of disease (Murray and Lopez 1996). One of the key elements of the burden of disease analysis is the quantification of human morbidity and mortality into costs of illness over the lifespan. The familiar disability-adjusted life years (DALYs) has allowed the estimation of the reduction in quality of life at the individual and societal level. Comparisons of the DALYs from different disease conditions may allow rational decisions about resource allocation. For example, in an African developing nation what might be the financial costs and reductions in DALYs attributable to community-level child deworming compared to the purchase of an advanced imaging machine in a capital city tertiary care hospital?

The second key insight from the global burden of disease studies is the growing importance of non-communicable diseases as a source of burden in developing nations (Murray and Lopez 1996). The face of illness in developing nations is no longer childhood preventable diseases. The success of the child survival activities of the 1970s and 1980s, especially oral rehydration and immunizations, has changed the age structure of these populations as many individuals live into their middle adult years (Jamison *et al.* 1993). Thus lifestyle diseases due to alcohol intake and cigarette smoking as well as obesity related metabolic and cardiovascular diseases and cancers are now exerting a major impact on illness, quality of life and mortality throughout the world. Combined with the chronic absolute and relative poverty in developing nations and lack of access to health care, and the weight of prevention activities still on, and properly so, child survival, these trends should raise serious concern for those of us interested in health and development.

These chapters also indicate the need for a critical socio-economic and political perspective in studies of health and disease. This is a new and important area of human population biology (Goodman and Leatherman 1998). At the same time scholars have turned to analysis of the impact of social and economic inequities on health (Kawachi *et al.* 1999). There is increasing evidence that the relative social and economic position in one's community and nation exerts powerful health effects. As biological anthropologists continue to focus their work in study populations experiencing economic development and modernization, we must pay attention not just to absolute income or wealth levels. The processes of development do not improve all members of society equally and at a similar rate. Certainly the maxim that the rich get richer is too simple, but it does contain the basic notion that the economic and political benefits of modernization remain out of the reach of the poorest and least educated parts of society.

Lastly, many of the chapters in this volume show that studies of health and disease must attempt a fine-grained study of lifestyle exposures. The new and re-emerging infections such as HIV/AIDS, tuberculosis, and the several parasitic diseases discussed here require detailed individual assessments of exposure as well as how these exposures are organized by social and economic structures discussed briefly above. These infectious diseases will be impossible to eradicate without full appreciation of the lifestyle behaviours associated with their transmission. Lifestyle factors also have a strong impact on some of the other diseases now becoming prevalent in society, such as Type 2 diabetes and also many of those which do not hit the headlines, but are incipient in daily life, as described in the chapter by Ford.

The opening chapter by Armelagos, provides the historical perspective to this topic by describing the changes seen in disease over time, both of patterns in existing disease and in the emergence of new diseases. It touches on the theme of urban health elucidated in more detail by Schell and Hills, and anticipates the treatment of emerging infections by Ford.

The second chapter by Ellison and Barrett focuses on contemporary human population biology by describing the life history perspective which seeks links and the interactions between disease states at different points in the human life cycle. An increased understanding of the evolution and transmission of disease over time and within the human life cycle has interest and future practical application for anthropologists, epidemiologists and biologists as well as health care planners. They correctly point out the opportunities to study fetal programming in developing country populations as economic development increases, the nutrition transition ensues and economic inequities increase. This will allow a thorough examination of the roles of poverty and related risk behaviours on growth, body size and metabolic parameters throughout the lifespan.

Mathematical modeling is a relatively new powerful research technique that, on the basis of existing data, can examine historical patterns of disease transmission and predict future changes in disease patterns and the spread of diseases. The chapter by Sattenspiel provides one such example using modelling to study the impact of population travel patterns on the spread of measles and ultimately a measles epidemic.

Section two of the Symposium volume looks at the current challenges. One of the more obvious challenges is measuring the impact of a person's genetic inheritance on their health. Williams-Blangero et al's thorough description of the concepts and techniques of genetic epidemiology to disentangle genetic effects on parasitic diseases is important as such diseases have persisted with an unchanging prevalence in spite of advances in medical science in the past 100 years. They provide excellent illustration of the impact of more simply defined genetic variation on measures of parasitic infection as well as the more interesting complex gene environment interactions. Lastly they provide good justification for the importance of genetic epidemiology in identifying susceptibility loci as part of rational drug discovery.

The impact of specific environmental factors, such as air, water and land pollutants has been thoroughly researched and there is clear documented evidence of an impact on health. Increasing environmental pollution is associated with developing and increasing urbanization and the chapter by Schell and Hills present the evidence for the health effects of urban pollution upon children. They also provide an important focus on the underlying demographic processes of urbanization, natural population growth and migration in the developing world.

Malaria in pregnancy is an enormous public health problem in Africa and Lindsay and coauthors present a very good review of the issues in the Gambia. Their emphasis on simple technology solutions is compelling. They also discuss the problems of malaria treatment in pregnancy reducing the development of immunity. More detailed research is required and experimental designs with ethically appropriate controls may be required to understand the multiple impacts on mother and child.

McGarvey et al report on a series of community-based studies in the Philippines and China to identify and treat cases of *Schistosoma japonicum*, another parasitic disease with a significant impact on population health. The described research projects may help refine more specifically our understanding of how partial immunity develops and its impact on infection and health, as well as allow general macro-level mathematical predictive modeling of transmission. This will be the only way to try to avert further *S. japonicum* transmission by interventions on agricultural, veterinary and water management activities.

Wareham attempts to unravel gene-environment interactions in Type 2 diabetes. This chapter emphasizes the candidate gene approach and case control designs. It is important to mention that many other types of designs are now being used to disentangle this complex disease, including various family designs and genome scans that may be quite efficient at identifying susceptibility loci and testing interactions with environmental exposures (Tsai *et al.* 2001).

HIV/AIDS is increasing globally and in Goodwin's chapter, based on research conducted in central and eastern Europe, the focus is on a biocultural perspective and showing key relationships between attitudes, values and behaviors and risk. Culturally based representations of individual values underlie behaviour. It is clear that an understanding of the factors that contribute to specific behaviours is essential in order for preventive interventions to be developed and implemented.

Changes in disease patterns and prevalence are as much to do with health care as with environmental, genetic and lifestyle issues. Kunitz tracks the changes in health profiles of the Indians of the United States with the decline in high levels of infectious diseases and emergence of increasing numbers of non-infectious conditions in parallel with the development of health care services for this population.

The chapter by Peters provides an overview of changes in disease over time mirrored by developments and change in public health and the Government's health care agenda. The author concludes that public health, in the broadest sense can address the changing face of disease. Finally, in the essay by Ford, the author summarizes the status quo with a strongly viewed perspective on human behaviour and disease.

This symposium volume contains presentations from the Society for the Study of Human Biology's 43rd symposium, which was jointly organized by the Society and the American Human Biology group for September 17th and 18th, 2001 in Cambridge, England. As readers will be aware, the events of September 11th, 2001 had repercussions throughout the world immediately after the event as well as in the longer term. The immediate consequences for this Symposium were that all flights out of the USA were cancelled and speakers and attendees from there were unable to get flights to England. Consequently some papers could not be delivered at the Symposium. This Symposium volume contains those papers not presented at the Symposium as well as those that were. It is therefore with pleasure that we offer this Symposium volume to readers and particularly to those attendees or potential attendees who missed either some or all of the programme because of events outside the control of the organizers.

References

Goodman, A.H. and Leatherman, T.L. (eds) (1998) *Building a New Biocultural Synthesis: Political-Economic Perspectives on Human Biology*: University of Michigan Press, Ann Arbor.

Jamison, D.T., Mosly, W.H., Measham, A.R. and Bobadilla, J.L. (eds) (1993) *Disease Control Priorities in Developing Countries*, New York: Oxford University Press.

Kawachi, I., Kennedy, B.P. and Wilkinson, R.G. (eds) (1999) 'Income Inequality and Health', vol. 1, *The Society And Population Health Reader*, New York: The New Press.

Murray, C.J.L., Lopez, A.D. (eds) (1996) *The Global Burden of Disease*, World Health Organization, Cambridge, MA: Harvard University Press.

Tsai, H-J., Sun, G., Weeks, D.E., Kaushal, R., Wolujewicz, M., McGarvey, S.T., Tufa, J., Viali, S. and Deka, R. (2001) 'Type 2 diabetes and three calpain-10 gene polymorphisms in Samoans: no evidence of association', *American Journal of Human Genetics*, 69: 1236–44.

Part I

Historical aspects

1 Emerging disease in the third epidemiological transition

George J. Armelagos

In 1969, William T. Stewart, the Surgeon General of the United States, testifying before Congress proposed that it was now 'time to close the book on infectious disease as a major health threat.' Stewart and others believed that with the development of antibiotics, vaccines and pesticides, we were on the verge of eradicating infectious disease. Buoyed by this success, Stewart's testimony before Congress was designed to position the United States public health system to meet its next health challenge: controlling chronic and degenerative diseases.

Stewart's assessment on the decline of infectious disease and the rise of chronic disease was the fulfillment of an epidemiological theory that was first proposed by Abdul Omran (1971), who argued that human populations were experiencing a shift in health and disease patterns. Omran contended that human disease history could be described as moving through a number of disease stages. Initially, humans passed through 'the age of pestilence and famine' to an 'age of receding pandemics' and finally into 'the age of degenerative and man-made diseases'. The basic feature of Omran's model (1971, 1977, 1983) was the idea that as infectious diseases were eliminated, chronic diseases would increase as the population aged. Finally, epidemiological transition theory had implications for demographic transition theory,[1] which suggested that after the decline in mortality there would be an eventual decline in fertility.

This chapter has three objectives; the first is to interpret and broaden the concept of epidemiological transition into a model that defines a number of dramatic shifts in disease patterns (Armelagos *et al.* 1996; Armelagos 1998; Armelagos and Barnes 1999; Barnes *et al.* 1999). Secondly, the evolution of emerging diseases will be discussed from the perspective of three epidemiological transitions (Barrett *et al.* 1998). While a distinct pattern of disease emerged as our Paleolithic ancestors moved into new ecological niches (Desowitz 1980), their mobility, small population size and low density precluded infectious disease from being a factor in the evolution of these populations. Finally, it will be shown that while emerging diseases have been a characteristic of human adaptation, following the shift to primary food production, there was an acceleration of the trend.

The traditional Hobbesian view of the 4,000,000 years of the Paleolithic was of the gatherer-hunters who foraged for their livelihood. Hobbes describes our ancestors living in 'continual fear' with 'a danger of violent death' and a life that was 'solitary, poor, nasty, brutish, and short' (*Leviathan*, i. xiii. 9). In actuality, Paleolithic populations appear to have been relatively healthy and well nourished. During the Neolithic, the shift to primary food production (agriculture) created the first epidemiological transition associated with the acceleration of emerging diseases. The second epidemiological transition (Omran's original epidemiological transition) began early in the 20th century with the decline in infectious disease and the rise of chronic diseases. We are entering the third epidemiological transition with the re-emergence of infectious diseases that were thought to be under control (many that are antibiotic resistant) and the rapid emergence of a number of 'new' diseases. The existence of antibiotic resistant pathogens (some that are resistant to multiple antibiotics) foretells a possibility that we are living in the eve of the antibiotic era. Finally, the third epidemiological transition is characterized by a transportation system, so vast, so rapid, that the globalization of the disease process (Waters 2001) has produced what has been called the 'viral superhighway'.

The concept of emerging disease needs to be considered more fully. Emerging infectious diseases are defined by the Institute of Medicine (IOM) as 'new, re-emerging, or drug-resistant infections whose incidence in humans has increased within the past decades or whose incidence threatens to increase in the near future' (Hughes 2001). For the public, emerging disease seems to have replaced the anxiety that was produced by the rampant fear of nuclear war following the Cold War of the 1950s. During that era, Hollywood produced a series of popular films such as *The Day the World Ended* (1956) in which radiation following a nuclear holocaust turns survivors into horrible mutants. Now, popular books such as *The Hot Zone* (Preston 1994) and movies such as *Outbreak* have captured the public's fascination with emerging diseases as threats to human survival. A monkey carrying a deadly new virus from central Africa infects and liquefies the organs of unwitting Californians, creating an epidemic that even the vast biomedical community is incapable of stopping, eventually threatening the annihilation of the human race. Just as we have begun to allay some of the fears of mutant pathogens running amok, there are serious concerns about the reality of bio-terrorism (McDade and Franz 1998; Henderson 1999; Arnon *et al.* 2001; Dennis 2001). The recent episode of anthrax in the United States attests to the impact that bio-terrorism has had on the western psyche.

Even the biomedical community's view on 'emerging' disease has been questioned. Paul Farmer (1996) suggests that emerging diseases are only 'discovered' when they have an impact on our daily existence. For example, Lyme disease (*Borrelia burgdorferi*) was studied long 'before suburban reforestation and golf courses complicated the equation by creating an environment agreeable to both ticks and affluent humans' (Farmer 1996). Even when a more holistic ecological perspective is taken, it is often limited to a position that considers 'emerging' disease as the result of human behavior or microbial changes that fails to place them in a broader political–economic context. There is a failure to 'ask how large-scale social forces influence unequally positioned individuals in increasingly interconnected populations' (Farmer 1996) and how inequality affects the disease process. Social inequalities are an essential element in understanding emerging disease patterns.

The world's biggest killer and greatest cause of ill health and suffering across the globe is listed almost at the end of the International Classification of Disease. It is given in code Z59.5 – extreme poverty (WHO 1995). The WHO ICD classification does not define *extreme* poverty. However, the need for basic human resources such as food, water, shelter, access to health care, and adequate social support, might allow us to make a minimal definition (Armelagos and Brown 2002). The World Bank estimates that three billion people in the world live on less than two dollars a day. Each year, in the developing world, 12.2 million children under five die. These deaths could be prevented for just a few cents per child. The World Bank's Global Burden of Disease Study pioneered the use of Disability Adjusted Life Years (DALYs) in assessing the impact of preventable disease (Hollinghurst *et al.* 2000) in the world today.

A perspective on the evolution of social inequality

The evolution of social inequality has not been studied extensively. There are only a few studies that analyze the gap that exists between individuals within a society and the gaps between societies. The relationship between inequality and health can be studied from an evolutionary perspective (Goodman *et al.* 1995). The analysis of changing patterns of health/disease and social organization in the prehistoric past allows us to better understand health inequalities in the contemporary world (Paynter 1989; McGuire and Paynter 1991).

Beginning in the Neolithic, inequalities within and between societies have accelerated with advances in technology (Figure 1.1). The gap between classes within society and differences in wealth among societies continues to widen. The disparity within and between nations in the present world order continues and the prospect that the gap will narrow is unlikely.

An evolutionary perspective allows us to examine the relationship between health and wealth, disease and poverty as part of continuing historical processes that

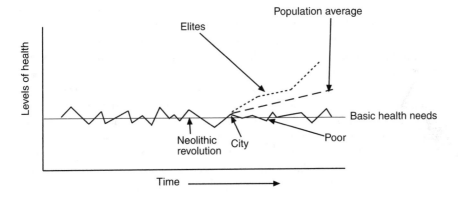

Figure 1.1 Cultural evolution, health and inequality.

have affected thousands of human generations. While stratification may occur without inequality, these quickly become inextricably linked. Social stratification originally evolved because it brought benefits to emerging elites which effected their well being. In most instances, the benefits accrued to the wealthy came at the expense of the poor (Armelagos and Brown 2002).

When evaluating the disease process, we tend to focus on pathogens (the microparasites) that use the host as a source of food and energy. However, just as microparasites are the source of disease, there are factors within societies that exacerbate survival to such an extent that they are as effective as the parasites themselves. For example, social stratification is an evolutionary strategy in which one segment exploits other segments of the social system to such a degree that their resources are limited and their health is at risk. The pattern and impact of exploitation is as parasitic as a pathogen. McNeil (1976) and Brown (1987) describe the process of exploitation as 'macroparasitism' and see it as a force in understanding the evolution of disease.

The evolution of technological change since the Neolithic has widened the gap between the rich and the poor, the healthy and the sick within and between societies. The gap between those at the top and the bottom of the social hierarchy, in the 21st century is greater than ever before in human history (Armelagos and Brown 2002).

Paleolithic baseline

The reconstruction of Paleolithic disease ecology requires the triangulation of methods and data from a number of sources. Archeological analysis of foraging populations provides direct evidence of their patterns of mortality and morbidity. The genomic diversity of pathogens and parasites provides clues to the phylogenetic relationships and patterns of adaptation to their hosts. Applying a molecular clock allows scientists to determine when the pathogen began to parasitize the host. Molecular analysis of the three modern taenid tapeworms that parasitize humans and were assumed to have become a problem during the Neolithic are now thought to have originated as human parasites in the Paleolithic (Hoberg *et al.* 2000, 2001).

The consideration of the disease ecology of contemporary gatherer-hunters provides a model for the types of disease that would have affected Paleolithic foragers. Sprent (1969a, 1969b) distinguishes two classes of parasites that would have afflicted gatherer-hunters. 'Heirloom species' are a class of parasites that have had a long-standing relationship with our anthropoid ancestors and that continued to infect them as they evolved to hominids. Head and body lice (*Pediculus humanus*), pinworms, and possibly yaws, malaria are heirloom species. Certain lice have been ectoparasites since the Oligocene (Laird 1989). Most of the internal protozoa found in modern humans and such bacteria as salmonella thyphi, and staphylococci (Cockburn 1967a, 1967b) are also heirloom species. In contrast to heirloom parasites that had longstanding relationships with anthropoids and hominids are 'souvenir' species that are 'picked up' during daily activity. Souvenir species are zoonoses whose primary hosts are non-human animals and they only incidentally infect humans.

Zoonoses are passed on to humans through insect bites, preparing and consuming contaminated flesh, and from animal bites. Sleeping sickness, tetanus, scrub typhus, relapsing fever, trichinosis, tularemia, avian or ichthyic tuberculosis, leptospirosis, and schistosomiasis are among the zoonotic diseases that likely afflicted earlier gatherer-hunters (Cockburn 1971). Small population size would have precluded infections of many bacteria and viruses. However, synanthropic relationships with the vectors served to maintain such human host-specific diseases as yellow fever and louse-borne relapsing fever (Laird 1989) in earlier foragers. Anopheles, the vector, necessary for transmission of malaria had evolved by the Miocene era by adapting to the canopy environment, suggesting that it would be present in the Paleolithic. Livingstone (1958) dismisses the threat of malaria in early hominids because of their small population size and an adaptation to the savanna, an environment that would not have included mosquitoes that carry the malaria plasmodium. If malaria was contracted, it would have been an isolated incident. Recent analysis of the genetic structure of variants of glucose-6-phosphate dehydrogenase confirms that malaria has only recently had a major impact on human populations (Tishkoff *et al.* 2001). The independent 'A' and 'Med' mutations in glucose-6-phosphate dehydrogenase suggest that this polymorphism originated at least 10,000 years ago.

The range of the earliest hominids was probably restricted to the tropical grassy woodland savannah, limiting the variety of pathogens that could be potential disease agents. As Dicke (1932) and Lambrecht (1964, 1967, 1980, 1985) note, hominids would have found extensive areas of Africa uninhabitable because of tsetse flies and the trypanosomes. Lambrecht also argues that as human species moved into new ecological niches the pattern of trypanosome infection would have changed. Beginning 200,000 years ago, as populations moved out Africa, there was expansion into temperate and tundra habitats that would have changed the disease ecology for the trypanosomes.

The diseases that are missing from the pantheon of Paleolithic pathogens are very informative. Contagious diseases such as influenza, measles, mumps, and smallpox would not have been present. There would have been few viruses infecting early hominids (Burnet 1962). Countering this claim, Cockburn (1967b) suggests that the viral diseases found in non-human primates would have been easily transmitted to early hominids.

Hominid populations remained stable throughout the Paleolithic. Fertility and mortality rates in populations would have to have been balanced for population size to remain low. Conventional wisdom has argued that Paleolithic populations experienced maximum fertility and high mortality. Armelagos *et al.* (1975) have offered an alternative scenario in the shift from gathering and hunting to agriculture. The picture that has emerged suggests a much bleaker picture of health. Instead of experiencing improved health, there is evidence of a substantial increase in infectious and nutritional disease following the shift to agriculture (Cohen and Armelagos 1984). The implication is that a population experiencing maximum fertility during the Paleolithic could not have increased fertility that would have led to population growth as their nutrition and health deteriorated.

The demographic changes following the Neolithic suggest that during the Paleolithic a stable population controlled by moderate fertility existed. Following the Neolithic revolution, there was a dramatic growth in population size and an increase in density even with a decline in nutrition and health because of increases in fertility.

The first epidemiological transition: disease in agricultural populations

The earliest evidence of primary food production in the Old World is from about 10,000 years ago. There are independent areas of cultivation in Mesopotamia (based on barley and wheat), sub-Saharan Africa (based on millets and plantains), Southeast Asia (based on rice), northern China (based on millet) and southern China (based on rice). Centres in the New World originate later based on the domestication of maize (Mesoamerica) and potatoes (South America). Significant settlements are evident in the Tigris–Euphrates area by 7000 years ago, and a thousand years later there are centralized governments controlling vast irrigation systems. This development created social classes with differential access to resources. In the Valley of Mexico, there were well-established settlements by 3500 BP and by 1 CE these settlements show extensive hierarchies.

Ecological changes increase the potential for disease load following the shift to primary food production. Sedentary villages increase parasitic disease infection by increasing contact with human waste. While sedentarism could and did occur prior to the Neolithic period in those areas with abundant resources, the shift to agriculture necessitated sedentary living. In sedentary populations the proximity of habitation areas and their waste deposit sites to the water supply is a source of contamination. In gathering-hunting groups, the frequent movement of the base camp and frequent forays away from the base camp by men and women, would decrease their contact with human wastes.

Animal husbandry also increased the frequency of contact with a steady supply of disease vectors. Zoonotic infections would have been contracted from domesticated animals, such as goats, sheep, cattle, pigs, and fowl, as well as the unwanted domestic animals such as rodents and sparrows, which developed permanent habitats in and around human dwellings. Products of domesticated animals such as milk, hair, and skin, as well as the dust raised by the animals, could transmit anthrax, Q fever, brucellosis, and tuberculosis. Breaking the sod during cultivation exposes workers to insect bites and diseases such as scrub typhus (Audy 1961). Livingstone (1958) showed that slash-and-burn agriculture in West Africa exposed populations to *Anopheles gambiae*, a mosquito that is the vector for *Plasmodium falciparum*, which causes malaria. The combination of disruptive environmental farming practices and the presence of domestic animals also increases human contact with arthropod vectors carrying yellow fever, trypanosomiasis, and filariasis, which now developed a preference for human blood. Some vectors developed dependent relationships with human habitats, the best example of which is *Aedes aegypti* (the vector for yellow fever and dengue), an artificial container breeder.

Various agricultural practices increased contact with non-vector parasites, such as irrigation (contact with schistosomal cercariae) and the use of faeces as fertilizer (infection with intestinal flukes) (Cockburn 1971).

The shift to agriculture heralded a change in ecology that resulted in diseases not frequently encountered by earlier foraging populations. The shift from a varied, well-balanced diet to one that contained fewer types of food sometimes resulted in dietary deficiencies. Food was stored in large quantities and widely distributed, probably resulting in outbreaks of food poisoning (Audy 1961). In Cohen and Armelagos (1984) there are a number of studies that show a decline in health following the Neolithic. The combination of a complex society, increasing divisions of class, epidemic disease, and dietary insufficiencies, no doubt added mental stress to the list of illnesses.

Urban development and disease

The growth of urban centres is a recent development in human history. In the Near East, large cities were established 6700 years ago. In the New World, large urban settlements were in existence by 1400 years ago. Urban centres at Memphis (Egypt) reached 30,000 souls by 3100 BCE, Ur (Babylonia) reached 65,000 inhabitants by 2030 BCE and Babylon had a population of 200,000 by 612 BCE (Chandler 1987). Settlements of this size increased the already difficult problem of removing human wastes and delivering uncontaminated water. Cholera, which is transmitted by contaminated water, was a potential problem. Diseases such as typhus (carried by lice) and the plague bacillus (transmitted by fleas or by the respiratory route), could be spread from person to person. Viral diseases such as measles, mumps, chicken-pox, and smallpox could be spread in a similar fashion. There were for the first time, during the period of urbanization, populations large enough to maintain disease in an endemic form. Cockburn (1967a) estimates that populations of one million would be necessary to maintain measles as an endemic disease. Others (Black *et al.* 1974) suggest that a population of only 200,000 would be required to maintain measles. Black and colleagues (1974) argue that a population of only a 1000 people is needed to sustain chicken-pox as an endemic disease. What was an endemic disease in one population could be the source of a serious epidemic disease in another group. Cross-continental trade and travel resulted in intense epidemics (McNeill 1976; Zinsser 1935). The Black Death took its toll in Europe in the 1300s. The epidemic eliminated at least a quarter of the European population (approximately 25 million people) (Laird 1989).

The period of urban development can also be characterized by the exploration and expansion of populations into new areas, which resulted in the introduction of novel diseases to groups that had little resistance to them (McNeill 1976). McNeill (1978) describes the process in which civilizations 'digest' the encountered populations as their disease vectors clear their path allowing easy access to expansion. For example, European-carried smallpox and measles destroyed millions of Native Americans following contact (Dobyns 1983; Ramenofsky 1987, 1993).

The exchange of disease can be a two-way street. For example, the exploration of the New World may have been the source of the treponemal infection that was transmitted to the Old World (Baker and Armelagos1988). The treponemal infection in the New World was endemic and not sexually transmitted (Rothschild *et al.* 2000). When introduced into the Old World there was a different mode of disease transmission. The sexual transmission of the treponeme created a different environment for the pathogen, and it resulted in a more severe and acute infection. Furthermore, crowding in the urban centres created changes in sexual practices, such as prostitution, and an increase in sexual promiscuity may have been a factor in the new venereal transmission of the pathogen (Hudson 1965). Claims that pre-Columbian syphilis existed in Europe have been made in response to the claims of New World origin of the disease (Pálfi *et al.*1992; Dutour *et al.* 1994). The resolution of this debate may await the recovery of material that can be identified as treponemal pathogen from Old World, pre-Columbian archaeological bone.

The process of industrialization, which began a little over 200 years ago, led to an even greater environmental and social transformation. London in 1800 was the only city in the world with a million inhabitants. City dwellers would be forced to contend with industrial wastes and polluted water and air. Slums that rose in industrial cities would become the focal point for poverty and the spread of disease. Epidemics of smallpox, typhus, typhoid, diphtheria, measles, and yellow fever in urban settings are well documented (Polgar 1964). Tuberculosis and respiratory diseases such as pneumonia and bronchitis are associated with harsh working situations and crowded living conditions. Urban population centres, with their extremely high mortality, were not able to maintain their population base through the reproductive capacity of those living in the city. Mortality outstripped fertility, requiring in-migration of rural populations to the city in order to maintain its numbers.

Recently much attention has been focused on the detrimental effects of industrialization on the international environment, including water, land, and atmosphere. Massive industrial production of commodities has caused pollution. Increasingly there is concern over the health implications of contaminated water supplies, over-use of pesticides in commercialized agriculture, atmospheric chemicals, and the future effects of depleted ozone on human health and food production. At no other time in human history have the changes in the environment been more rapid and so extreme. Increasing incidence of cancer among young people and the increase in respiratory disease has been implicated in these environmental changes.

The United Nations Population Fund (2001) reports that in 2000, 47 per cent (2.9 billion people) of the world population are living in an urban setting. In 30 years, that number will increase to 60 per cent. WHO has addressed the issue of health in urban settings with their healthy cities initiative (Goldstein 2000; Kenzer 2000; Tsouros 2000). The WHO programme attempts to systematically address poverty, the vulnerability of segments of the populations, and the lack of access that these populations have to health care.

The second epidemiological transition: the rise of chronic and degenerative disease

Traditionally, the term 'epidemiological transition' refers to the shift from acute infectious diseases to chronic non-infectious, degenerative diseases. The increasing prevalence of these chronic diseases is related to an increase in longevity. Cultural advances result in a larger percentage of individuals reaching the older age segments of the population. In addition, the technological advances that characterize the second epidemiological transition result in an increase in environmental degradation. An interesting characteristic of many of the chronic diseases is that they are particularly prevalent and 'epidemic-like' in transitional societies, or in those populations undergoing the shift from developing to developed modes of production. In developing countries, many of the chronic diseases associated with the epidemiological transition appear first in members of the upper socioeconomic strata (Burkitt 1973), because of their access to Western products and practices.

With increasing developments in technology, medicine, and science, the germ theory of disease causation developed. While there is some controversy as to the role that medicine played in the decline of some of the infectious diseases (McKeown 1979), there was a better understanding of the source of infectious disease and this admittedly resulted in increasing control over many infectious diseases. The development of immunization resulted in the control of many infections and recently was the primary factor in the eradication of smallpox. In the developed nations, a number of other communicable diseases have diminished in importance. The decrease in infectious diseases and the subsequent reduction in infant mortality have resulted in greater life expectancy at birth. The increase in longevity for adults has resulted in an increase in chronic and degenerative diseases.

Many diseases of the second epidemiological transition share common, etiological factors related to human adaptation, including diet, activity level, mental stress, behavioural practices, and environmental pollution. For example, the industrialization and commercialization of food often results in malnutrition, especially for those societies in 'transition' from subsistence forms of food provision to agri-business. Many do not have the economic capacity to purchase food that meets their nutritional requirements (Fleuret and Fleuret 1980). Obesity and high intakes of refined carbohydrates are related to the increasing incidence of heart disease and diabetes. Obesity is considered to be a common form of malnutrition in developed countries and is a direct result of an increasingly sedentary life-style in conjunction with steady or increasing caloric intakes. A unique characteristic of the chronic diseases is their relatively recent appearance in human history as a major cause of morbidity. According to Corruccini and Samvit (1983), this is indicative of a strong environmental factor in disease etiology. While biological factors such as genetics are no doubt important in determining who is most likely to succumb to which disease, genetics alone cannot explain the rapid increase in chronic disease. Critics of McKeown have focused on his use of evidence for improved nutrition (Johansson 1991, 1992; Schofield and Reher 1991) and failure to consider improvements in public health practices (Woods 1990; Johansson 1991, 1992; Kunitz 1991; Schofield and Reher 1991).

The third epidemiological transition

Human populations are in the midst of the third epidemiological transition. There is a re-emergence of infectious diseases that have multiple antibiotic resistance. Furthermore, the emergence of diseases has a potential for having a global impact. In a sense, the contemporary transition does not eliminate the possible co-existence of infectious diseases typical of the first epidemiological transition (some 10,000 years ago) in our own time; the World Health Organization reports that of the 50,000,000 deaths each year, 17,500,000 are the result of infectious and parasitic disease. WHO states that two billion people in the world are infected with hepatitis B virus (WHO 1995). Two billion of the world's population have tuberculosis (8 million cases contracted every year and 3 million die in that period). In the last thirty years, 40 million people have become infected with HIV and 3 million people have died during that period.

Humans have lived in urban centres for only 0.125 per cent of our history. This may explain the paucity of evidence for the genetic response to specific disease. Svanborg-Eden and Levin (1990) challenge the proposition that infectious disease is a major force in the selection and evolution of genetic variability in human populations. They argue that there are three constraints for infectious disease to act as an effective agent of natural selection. First, most variation in the frequency of infectious disease is the result of environmental factors. Second, the array of host defenses is general in their actions and overlapping in their functions. Third, the specific immune defences are adaptive at the somatic level and therefore there is less of a need for selection leading to germ-line evolution. Four decades ago, Lederberg (1963) suggested that diseases that had animal reservoirs could lead to the development of disease resistance in human populations. He argued that the persistence of 'small differentials' could lead to genetic immunity.

The re-emergence of infectious diseases has been one of the most interesting evolutionary stories of the last decade and has captured the interest of scientists and the public. Satcher (1995) and Lederberg (1997) list almost thirty diseases that have emerged in the last 28 years. A list of the most recent emerging diseases include Rotavirus (1973), Parvovirus B19 (1975), *Cryptosporidium parvum* (1976), Ebola virus (1977), *Legionella pneumophila* (1977), Hantaan virus (1977), *Campylobacter jejuni* (1977), HTLV I (1980), *Staphylococcus aureus* toxin (1981), Escherichia coli 0157:h7 (1982), HTLV II (1982), *Borrelia burgdorferi* (1982), HIV (1983), *Helicobacter pylori* (1983), *Enterocytozoon bieneusi* (1985), Human Herpes-virus-6 (1988), Hepatitis E (1988), *Ehrlichia chafeensis* (1989), Hepatitis C (1989), Guanarito virus (1991), *Encephalitozoon hellen* (1991), New species of *Babesia* (1991), *Vibrio Cholera* 0139 (1992), *Bartonella (=Rochalimaea) henselea* (1992), Sin Nombre virus (1993), *Encephalitozoon cuniculi* (1993), Sabia virus (1994), and HHV-8 (1995).

The Institute of Medicine (IOM) (Lederberg *et al.* 1992) reports that the emergence of disease is the result of an interaction of social, demographic, and environmental changes in a global ecology and in the adaptation and genetics of the microbe. Similarly, Morse (1995) sees emerging disease as a result of demographic changes, international commerce and travel, technological change, breakdown of public health measures, and microbial adaptation.

Among the ecological changes Morse describes are agricultural development projects, dams, deforestation, floods, droughts and climatic changes that resulted in the emergence of diseases such as Argentine hemorrhagic fever, Korean hemorrhagic fever (Hantaan) and Hantavirus pulmonary syndrome. Human demographic behaviour has been a factor in the spread of dengue, the source for the introduction and spread of HIV and other sexually transmitted disease.

The engine that is driving the re-emergence of many of these diseases is the ecological change that brings humans into contact with pathogens. Except for Brazilian pururic fever which may represent a new strain of *Haemophilus influenzae, biotype aegyptius* most of the emerging diseases are of cultural origin. The role of humans in the development of antibiotic resistance by medical and agricultural practices is well established. Humans are clearly 'the world's greatest evolutionary force' (Palumbi 2001). Palumbi argues:

> Human ecological impact has enormous evolutionary consequences as well and can greatly accelerate evolutionary change in the species around us, especially disease organisms, agricultural pests, commensals, and species hunted commercially. For example, some forms of bacterial infection are insensitive to all but the most powerful antibiotics, yet these infections are increasingly common in hospitals. Some insects are tolerant of so many different insecticides that chemical control is useless. Such examples illustrate the pervasive intersection of biological evolution with human life, effects that generate substantial daily impacts and produce increasing economic burden.
>
> (Palumbi 2001)

The acceleration of evolution, according to Palumbi, costs the United States at least $33 billion and as much as $50 billion a year in costs related to the antibiotic and pesticide resistance in organisms.

Conclusion: 'getting ahead of the curve'

The CDC (1994), following the IOM recommendation, proposed a plan, which was later modified (CDC 1998), suggesting a four-pronged attack on emerging disease. First, they emphasize a need to strengthen infectious surveillance and response that can detect and contain infectious agents. Second, they address the need to research issues raised by these challenges. Third, there is a need to repair the public health infrastructures and training. Finally, there is a need to strengthen prevention and control programs 'locally, nationally, and globally'. The CDC and IOM plans focus on detection and response that enhances the capacity of infrastructure to detect and respond to disease threats.

David Satcher (1995), Director of the CDC, in the inaugural issue of *Emerging Infectious Disease* is forceful in his assessment of what may be the key in responding to infectious diseases:

> We cannot overstate the role of behavioral science in our effort to get ahead of the curve with emerging infections. Having the science or laboratory technology to control infectious diseases is not enough, unless we can influence people to behave in ways that minimize the transmission of infections and maximize the efforts of medical interventions. For example, even though HIV/AIDS does not have a vaccine or cure, it is almost entirely preventable. For many people, however, reducing the risk for HIV infection and AIDS requires important changes in lifestyle or behavior. We must use our knowledge of human behavior to help people make lifestyle changes and prevent disease.
>
> (Satcher 1995)

We understand the proximate and ultimate causes of disease. We understand the cultural practices that allow the pathogen to 'jump' their species barrier and escape their geographic boundaries. The observation that the widening economic and political gap has been the pattern of our post-Neolithic history does not mean that it is inevitable. The issue of inequality must be addressed. The question remains as to our will to deal with inequality. To date, the issue seems to have been avoided and accepted as an inevitable aspect of evolution. It is time to reconsider the issue.

These are the words with which Edward B. Tylor (1881) chose to close his book *Anthropology*, written over a century ago:

> Readers who have come thus far need not be told in so many words of what the facts must already brought to their minds – that the study of man and civilization is not only a matter of scientific interest, but at once passes into the practical business of life. We have in it the means of understanding our lives in and our place in the world, vaguely and imperfectly it is true, but at any rate more clearly than any former generation. The knowledge of man's course of life, from remote past to the present, will not only help us forecast the future, but may guide us in our duty of leaving the world better than we found it.
>
> (Tylor 1881: 439–40)

The words echo advice that we can live with today.[2]

Notes

1 Demographic transition theory is a generalized model of population structure that is the basis for understanding fertility and mortality processes in contemporary populations. In the first stage, which was thought to represent most of human evolutionary history, populations were at their highest natural fertility and high mortality, resulting in little increase in natural population. In the second stage, there is a decrease in mortality and natural fertility remains high, resulting in a rapid increase in population. The third stage, mortality rates are low and birthrates begin to decline, resulting in a slow population growth. In the last stage, there is low fertility and low

mortality with no natural increase in population. Given the evidence that Paleolithic populations were controlling fertility, demographic transition theory needs to be re-evaluated.

2 I would like to thank Lynn M. Sibley, Bethany Turner and Diana Smay for thoughtful comments on earlier versions of this paper.

References

Armelagos, G.J. (1998) 'The viral superhighway', *The Sciences*, 38: 24–30.

Armelagos, G.J. and Barnes, K. (1999) 'The evolution of human disease and the rise of allergy: epidemiological transitions', *Medical Anthropology*, 18: 187–213.

Armelagos, G.J. and Brown, P.J. (2002) 'The body of evidence', in R. Steckels and J. Rose (eds) *The Backbone of History*, Cambridge: Cambridge University Press.

Armelagos, G.J., Goodman, A.H. and Jacobs, K. (1975) 'The origins of agriculture: population growth during a period of declining health', in W. Hern (ed.) 'Cultural change and population growth: an evolutionary perspective', *Population and Environment* 13(1): 9–22.

Armelagos, G.J., Barnes, K.C. and Lin, J. (1996) 'Disease in human evolution: the re-emergence of infectious disease in the third epidemiological transition', *AnthroNotes*, 18: 1–7.

Arnon, S.S., Schechter, R., Inglesby, T.V., Henderson, D.A., Bartlett, J.G., Ascher, M.S., Eitzen, E., Fine, A.D., Hauer, J., Layton, M., Lillibridge, S., Osterholm, M.T., O'Toole, T.O., Parker, G., Perl, T.M., Russell, P.K., Swerdlow, D.L. and Tonat, K. (2001) 'Botulinum toxin as a biological weapon: medical and public health management', *JAMA*, 285: 1059–70.

Audy, J.R. (1961) 'The ecology of scrub typhus', in J.M. May (ed.) *Studies in Disease Ecology*, pp. 389–432, Studies in Medical Geography, New York: Hafner Publishing.

Baker, B. and Armelagos, G.J. (1988) 'Origin and antiquity of syphilis: a dilemma in paleopathological diagnosis and interpretation', *Current Anthropology*, 29(5): 703–37.

Barnes, K.C., Armelagos, G.J. and Morreale, S.C. (1999) 'Darwinian medicine and the emergence of allergy', in W. Trevethan, J.McKenna, and E.O. Smith (eds) *Evolutionary Medicine*, New York: Oxford University Press.

Barrett, R., Kuzawa, C.W., McDade, T. and Armelagos, G.J. (1998) 'Emerging infectious disease and the third epidemiological transition' in W. Durham (ed.) *Annual Review Anthropology*, vol. 27, pp. 247–71, Palo Alto: Annual Reviews Inc.

Black, F.L., Hierholzer, W.J., Pinheiro, F., Evans, J.A.S., Woodall, J.P., Opton, E.M., Emmons, J.E., West, B.S., Edsall, G., Downs, W.G. and Wallace, G.D. (1974) 'Evidence for persistence of infectious agents in isolated human populations', *American Journal of Epidemiology*, 100: 230–50.

Brown, P.J. (1987) 'Microparasites and macroparasites', *Cultural Anthropology*, 2: 155–71.

Burkitt, D.P. (1973) 'Some disease characteristics of modern western medicine', *British Medical Journal*, 1: 274–8.

Burnet, F.M. (1962) *Natural History of Infectious Disease*, Cambridge: Cambridge University Press.

CDC (1994) *Addressing Emerging Infectious Disease Threats: A Prevention Strategy for the United States*, Atlanta: U.S. Public Health Service.

CDC (1998) *Preventing Emerging Infectious Diseases: A Strategy for the 21st Century*, Atlanta: U.S. Public Health Service.

Chandler, T. (1987) *Four Thousand Years of Urban Growth: An Historical Census*, Lewiston, NY: Edward Mellen Press.

Cockburn, T.A. (1967a) 'The evolution of human infectious diseases', in T.A. Cockburn (ed.) *Infectious Diseases: Their Evolution and Eradication*, pp. 84–107, Springfield, IL: Charles C. Thomas.

Cockburn, T.A. (1967b) 'Infections of the order primates', in T.A. Cockburn (ed.) *Infectious Diseases: Their Evolution and Eradication*, Springfield, IL: Charles C. Thomas.

Cockburn, T.A. (1971) 'Infectious disease in ancient populations', *Current Anthropology*, 12: 45–62.

Cohen, M.N. and Armelagos, G.J. (eds) (1984) *Paleopathology at the Origins of Agriculture*, New York: Academic Press.

Corruccini, R.S. and Samvit, S.K. (1983) 'The epidemiological transition and the anthropology of minor chronic non-infectious diseases', *Medical Anthropology*, 7: 36–50.

Dennis, C. (2001) 'The bugs of war', *Nature*, 411: 232–5.

Desowitz, R.S. (1980) 'Epidemiological–ecological interactions in savanna environments', in D.R. Harris (ed.) *Human Ecology in Savanna Environments*, pp. 457–77, London: Academic Press.

Dicke, B.H. (1932) 'The tseste fly's influence on South African history', *South African Journal of Science*, 29: 792.

Dobyns, H. (1983) *Their Numbers Become Thinned: Native American Population Dynamics in Eastern United States*, Knoxville: University of Tennessee Press.

Dutour, O., Pálfi, G., Bérato, J. and Brun, J.-P. (1994) *L'origine de la syphilis en Europe – avant ou aprés 1493?*, Paris: Errance.

Farmer, P. (1996) 'Social inequalities and emerging infectious diseases', *Emerging Infectious Diseases*, 2: 259–69.

Fleuret, P. and Fleuret, A. (1980) 'Nutrition, consumption and agricultural change', *Human Organization*, 39: 250–60.

Goldstein, G. (2000) 'Healthy cities: overview of a WHO international program', *Reviews on Environmental Health*, 15: 207–14.

Goodman, A.H., Martin, D.L. and Armelagos, G.J. (1995) 'The biological consequences of inequality in prehistory', *Rivista di Antropologia (Roma)* 73: 123–31.

Henderson, D.A. (1999) 'The looming threat of bioterrorism', *Science*, 283: 1279–82.

Hoberg, E.P., Jones, A., Rausch, R.L., Eom, K.S. and Gardner, S.L. (2000) 'A phylogenetic hypothesis for species of the genus Taenia (Eucestoda : Taeniidae)', *Journal of Parasitology*, 86: 89–98.

Hoberg, E.P., Alkire, N.L., Queiroz, A. de and Jones, A. (2001) 'Out of Africa: origins of the Taenia tapeworms in humans', *Proceedings of the Royal Society of London – Series B: Biological Sciences*, 268: 781–7.

Hollinghurst, S., Bevan, G. and Bowie, C. (2000) 'Estimating the "avoidable" burden of disease by disability adjusted life years (DALYs)', *Health Care Management Science*, 3: 9–21.

Hudson, E.H. (1965) 'Treponematosis and man's social evolution', *American Anthropologist*, 67: 885–901.

Hughes, J.M. (2001) 'Emerging infectious diseases: a CDC perspective', *Emerging Infectious Diseases*, 7(3): 494–6.

Johansson, S.R. (1991) 'The health transition: the cultural inflation of morbidity during the decline of mortality', *Health Transition Review*, 1: 39–68.

Johansson, S.R. (1992) 'Measuring the cultural inflation of morbidity during the decline in mortality', *Health Transition Review*, 2: 78–89.

Kenzer, M. (2000) 'Healthy cities: a guide to the literature', *Public Health Reports*, 115: 279–89.

Kunitz, S.J. (1991) 'The personal physician and the decline of mortality', in D.R.R. Sclofield and A. Bideau (ed.) *The Decline of Mortality in Europe*, pp. 248–62, Oxford: Clarendon Press.

Laird, M. (1989) 'Vector-borne disease introduced into new areas due to human movements: a historical perspective', in M.W. Service (ed.) *Demography and Vector-Borne Diseases*, pp. 17–33, Boca Raton, FL: CRC Press.

Lambrecht, F.L. (1964) 'Aspects of evolution and ecology of Tsetse flies and Trypanosomaisis in prehistoric African environments', *Journal of African History*, 5: 1–24.

Lambrecht, F.L. (1967) 'Trypanosomiasis in prehistoric and later human populations: A tentative reconstruction', in D. Brothwell and A.T. Sandison (eds) *Diseases in Antiquity*, pp. 132–51, Springfield, IL: Thomas.

Lambrecht, F.L. (1980) 'Paleoecology of tsetse flies and sleeping sickness in Africa', *Proceedings of the American Philosophical Society*, 124: 367–385.

Lambrecht, F.L. (1985) 'Trypanosomes and hominid evolution', *Bioscience*, 35: 640–6.

Lederberg, J. (1963) 'Comments on A. Motulsky's genetic systems in disease susceptibility in mammals', in W.J. Schull (ed.) *Genetic Selection in Man*, pp. 112–260 (interspersed with Motulsky text). Ann Arbor: University of Michigan Press.

Lederberg, J. (1997) 'Infectious disease as an evolutionary paradigm', *Emerging Infectious Diseases*, 3(4): 417–23.

Lederberg, J., Shope, R.E. and Oaks, S.C. (eds) (1992) *Emerging Infection: Microbal Threats to Health in the United States*, Institute of Medicine, National Academy Press.

Livingstone, F.B. (1958) 'Anthropological implications of sickle-cell distribution in West Africa', *American Anthropology*, 60: 533–62.

McDade, J.E. and Franz, D. (1998) 'Bioterrorism as a public health threat', *Emerging Infectious Diseases*, 4: 493–4.

McGuire, R. and Paynter, R. (eds) (1991) *The Archaeology of Inequality*, Oxford: Basil Blackwell.

McKeown, T. (1979) *The Role of Medicine: Dream, Mirage or Nemisis*, Princeton: Princeton University Press.

McNeill, W.H. (1976) *Plagues and People*, Garden City: Anchor/Doubleday.

McNeill, W.H. (1978) 'Disease in history', *Social Science and Medicine*, 12: 79–81.

Morse, S.S. (1995) 'Factors in the emergence of infectious diseases', *Emerging Infectious Diseases*, 1: 7–15

Omran, A.R. (1971) 'The epidemiologic transition: a theory of the epidemiology of population change', *Millbank Memorial Fund Quarterly*, 49: 509–37.

Omran, A.R. (1977) 'A century of epidemiologic transition in the United States', *Preventive Medicine*, 6: 30–51.

Omran, A.R. (1983) 'The epidemiologic transition theory: a preliminary update', *Journal of Tropical Pediatrics*, 29: 305–16.

Pálfi, G., Dutour, O., Borréani, M., Brun, J.-P. and Bérato, J. (1992) 'Pre-Columbian congenital syphilis from the Late Antiquity in France', *International Journal of Osteoarcheology*, 2: 245–261.

Palumbi, S.R. (2001) 'Humans as the world's greatest evolutionary force', *Science*, 293: 1786–90.

Paynter, R. (1989) 'The archaeology of equality', *Annual Review of Anthropology*, 18: 369–99.

Polgar, S. (1964) 'Evolution and the ills of mankind' in S. Tax (ed.) *Horizons of Anthropology*, pp. 200–211, Chicago: Aldine.

Preston, R. (1994) *The Hot Zone*, New York: Random House.

Ramenofsky, A. (1987) *Vectors of Death: The Archaeology of European Contact*, Albuquerque, NM: University of New Mexico Press in association with the Center for Documentary Studies at Duke University.

Ramenofsky, A. (1993) 'Diseases in the Americas, 1492–1700', in K. Kiple (ed.) *The Cambridge World History of Human Disease*, New York: Cambridge University Press.

Rothschild, B.M., Calderon, F.L., Coppa, A. and Rothschild, C. (2000) 'First European exposure to syphilis: the Dominican Republic at the time of Columbian contact', *Clinical Infectious Diseases*, 31: 936–41.

Satcher, D. (1995) 'Emerging infections: getting ahead of the curve', *Emerging Infectious Diseases*, 1(1): 1–6.

Schofield, R. and Reher, D. (1991) 'The decline of mortaility in Europe', in R. Schofield, D. Reher and A. Bideau (eds) *The Decline of Mortality in Europe*, pp. 1–17, Oxford: Clarendon Press.

Sprent, J.F.A. (1969a) 'Evolutionary aspects of immunity of zooparasitic infections', in G.J. Jackson (ed.) *Immunity to Parasitic Animals*, vol. 1, pp. 3–64, New York: Appleton.

Sprent, J.F.A. (1969b) 'Helminth "zoonoses": an analysis', *Helminthol. Abstr.* 38: 333–51.

Svanborg-Eden, C. and Levin, B.R. (1990) 'Infectious disease and natural selection in human populations', in A.C. Swedlund and G.J. Armelagos (eds) *Disease in Populations in Transition*, pp. 31–48, New York: Bergfin and Garvey.

Tishkoff, S.A., Varkonyi, R., Cahinhinan, N., Abbes, S., Argyropoulos, G., Destro-Bisol, G., Drousiotou, A., Dangerfield, B., Lefranc, G., Loiselet, J., Piro, A., Stoneking, M., Tagarelli, A., Tagarelli, G., Touma, E.H., Williams, S.M. and Clark, A.G. (2001) 'Haplotype diversity and linkage disequilibrium at human G6PD: recent origin of alleles that confer malarial resistance', *Science*, 293: 455–62.

Tsouros, A.D. (2000) 'Why urban health cannot be ignored: the way forward', *Reviews on Environmental Health*, 15: 267–71.

Tylor, E.B. (1881) *Anthropology: An Introduction to the Study of Man and Civilization*, New York: D. Appleton.

United Nations Population Fund (2001) *Development Levels and Environmental Impact: State of World Population 2001*, New York: The United Nations Population Fund.

Waters, W.F. (2001) 'Globalization, socioeconomic restructuring, and community health', *Journal of Community Health*, 26: 79–92.

WHO (1995) *Executive Summary. The World Health Report: Bridging the Gaps*, Geneva: World Health Organization.

Woods, R. (1990) 'The role of public health in the nineteenth-century mortality decline', in J. Caldwell, S. Findley, P. Caldwell, G. Santow, W. Cosford, J. Braid and D. Broers-Freeman (eds), *What We Know About Health Transition: The Cultural, Social, and Behavioral Determinants of Health*, pp. 110–15, Proceedings of an International Workshop, vol. 1, Canberra: Health Transition Centre.

Zinsser, H. (1935) *Rats, Lice and History*, Boston: Little, Brown.

2 Life history perspectives on human disease

Peter T. Ellison and Emily S. Barrett

Introduction

Life historical approaches to disease are attracting increasing attention within both epidemiology and evolutionary biology. The concept of life history refers to the interactions and linkages between the states of an organism at different points in its life cycle. Although an appreciation for these linkages has long been implicit within both epidemiology and evolutionary biology, the development of empirical approaches on the one hand and theoretical approaches on the other has presented formidable difficulties in data collection, analysis, and modeling. As these difficulties have been overcome, new data, theory, and speculation have rapidly accumulated.

The development of life historical approaches within epidemiology was made possible by systems of record linkages. These allow researchers to join information on disease morbidity to information on reproductive history, pre- and perinatal conditions, and familial variables available from birth records and other sources. With these linked data sets comes the ability to generate longitudinal data for individuals. Replacing cross-sectional data with linked, longitudinal data can produce very different observations and conclusions. For example, Bakketeig and Hoffman (1981) were able to demonstrate using Norwegian data that the apparent U-shaped relationship between parity and frequency of pre-term birth is an artifact of cross-sectional analysis. Substituting longitudinal data and grouping by completed family size demonstrated instead that the frequency of pre-term birth declines monotonically with parity for all levels of achieved fertility (Figure 2.1). Grouping individuals by completed family size is obviously only possible when information from later in the life cycle can be linked to the incidence of pre-term birth at an earlier stage in the life cycle.

The principal use of life historical data within epidemiology is not to correct the distortions of cross-sectional analysis, however, but to help discriminate the relative importance of genetic, developmental, and acute environmental sources of variance in morbidity and disease outcomes. In the absence of longitudinal data, only concurrent states of the organism and its environment can be analyzed as sources of this variance. This can lead to a confounding of static variables (e.g. growth status, nutritional status, reproductive status) with dynamic variables (e.g. rate of growth, energy balance, reproductive history), and so to an underestimate

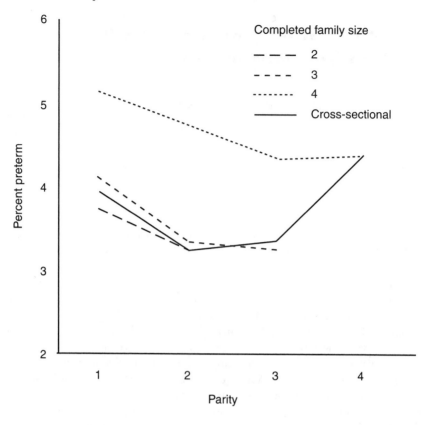

Figure 2.1 The percentage of pre-term births by parity in a Norwegian population. Considered cross-sectionally the percentage of pre-term births displays a U-shaped relationship to parity. Calculated longitudinally and grouped by completed family size, the percentage of pre-term births shows a monotonic decrease with increasing parity for all groups. (Data from Bakketeig and Hoffman 1981.)

of the importance of developmental correlates of disease. It should be noted, however, that confusions of this nature are by no means eliminated simply by employing longitudinal data. For example, recent re-analysis of longitudinal British data indicates that the previously reported relationship between birth weight and age at menarche, which persists when one controls growth *status*, disappears when one controls growth *rate* (dos Santos Silva *et al.* 2002).

The life historical perspective has also long been integral to evolutionary biology and has recently produced an efflorescence of both theoretical and empirical research. A key feature of the life historical perspective in evolutionary biology, particularly as applied to disease, is the concept of trade-offs. In an early, influential paper, Gadgil and Bossert (1970) posited a simple trade-off in the allocation of energy and other resources between the physiological categories of growth, maintenance, and reproduction. This model, which has served as the foundation for research in this area, is of course a drastic simplification, collapsing many levels of trade-offs within each of the major categories. For example, we can easily identify trade-offs between allocations to different tissues within the category of growth; trade-offs between different aspects of immune function within the category of maintenance; and trade-offs between parenting effort and mating effort within the category of reproduction. Not only are there trade-offs between different physiological categories at any point in time, there are trade-offs between outcomes at different points in the life cycle. Positive outcomes at one point in time can be linked to negative outcomes at another with the net cost or benefit to the organism only being calculable across the entire life cycle.

In evolutionary biology the life historical perspective is often employed to identify optimal trajectories through the branching forest of trade-offs that comprise an individual life cycle. Where empirical research identifies life historical patterns and associations, these are assumed in the first instance to be functional, shaped by natural selection to serve the reproductive fitness of the organism. It should be noted that this emphasis on the functionality of life historical patterns is not explicitly shared by the epidemiological perspective. Life historical patterns associated with disease outcomes may presumptively represent varieties of pathology for epidemiologists, even as disease itself is presumed to be pathological. For the evolutionary biologist, life historical patterns associated with disease outcomes may still be viewed as functional, the disease outcome merely representing the cost of some trade-off with a greater associated benefit elsewhere in the life cycle. Williams (1966) and Hamilton (1966) originally employed this logic to explain how senescence itself can be viewed as the product of life historical trade-offs that are ultimately functional and adaptive.

Although their orientations toward pathology and functionality may be different, the epidemiological and evolutionary approaches to life history and disease are largely complementary and together can illuminate many features of the changing landscape of human disease. This chapter illustrates the value of combining these perspectives by briefly reviewing three current examples of life historical approaches to disease. Examples have been chosen that can be associated with each of Gadgil and Bossert's three major categories of energy allocation: growth, maintenance, and reproduction. In reviewing these examples attention will be drawn to the importance of longitudinal data and to the heuristic value of the concept of trade-offs in energy allocation and the functional optimization of developmental trajectories.

Fetal growth and adult disease

The Barker hypothesis, also referred to as the fetal origins hypothesis or fetal programming, suggests that many of today's leading health concerns can be attributed largely to prenatal and early childhood influences (Barker 1994). According to this hypothesis, an individual's early environment can exert strong effects on the body's development, resulting in permanent changes in structure and function. Proponents of the hypothesis argue that altered perinatal growth underlies predispositions to a range of adult illnesses, including coronary heart disease, diabetes, stroke, and high blood pressure.

Underlying this hypothesis is the idea that if energetic intake is limited during development, not all organ systems can develop in an optimal fashion. Trade-offs may be made which involve sustaining the growth and function of certain organ systems at the expense of others. In the case of humans, brain tissue is so meta-bolically expensive, yet so crucial to fitness, that when resources are limited, selection may have favoured the diversion of energy from somatic systems in order to maintain proper brain development. Somatic organ structure (and therefore function) may be affected in such individuals, as oxygen or nutrient starved cells stimulate a hormonally mediated decrease in cell division. As a result, not only may the affected organs have abnormal numbers and distributions of cells, but irregular hormone secretion and metabolic activity as well. On a macroscopic level, these abnormalities may be manifest as slowed fetal growth and therefore low birth weight, reduced length, a low ponderal index (birthweight/length^3), or a disproportionately small abdominal circumference (Barker 1995). Fetal genes mediating this prenatal response to suboptimal conditions would be selected for even if they led to health problems or premature death in an individual's post-reproductive life.

Perhaps the most well-documented correlation that has emerged from empirical studies of longitudinal data links infant birth weight and later risk of coronary heart disease (Figure 2.2). A number of studies have demonstrated an inverse relationship between the two; that is, individuals with the lowest birth weights have the highest rates of death from heart disease as adults (Barker *et al.* 1989; Leon *et al.* 1998) and are also more likely to develop hypertension, high cholesterol, and insulin-related disorders (Godfrey and Barker 2000). Measures of small size at birth are generally interpreted as indicative of slow rates of growth *in utero*, which, in turn, are taken as a proxy for fetal undernutrition. The effects of retarded fetal growth may be further compounded during early childhood development. Infants who are of low weight at age one appear to run high risks of developing heart disease as adults (Fall *et al.* 1995). Yet it has been reported that the group at greatest risk for heart disease later in life is that of individuals who had low birth weights but subsequently caught up to have average or greater than average body mass by age seven (Eriksson *et al.* 1999; Figure 2.3). This suggests that overnutrition later in life is particularly incompatible with the physiology characteristic of fetal undernourishment. The particular cellular and biochemical mechanisms implicated, however, have yet to be identified.

More specific mechanisms have been hypothesized to explain the link between Type 2 diabetes (or more broadly, 'insulin resistance syndrome') and thinness at birth (Figure 2.4). A thin fetus is likely to have relatively little muscle tissue, which would normally be a primary peripheral site of insulin action. Under conditions of restricted energy availability, muscles may become insulin resistant to ensure sufficient energy allocation to brain tissue development (Barker 1995). Supporting evidence indicates that the fetal insulin production necessary for early growth is reduced in infants with low birth weights, low placental weights, or relatively thin abdomens (Godfrey *et al.* 1996). Retarded fetal growth may then result in a resetting of the HPA (hypothalamus-pituitary-adrenal) axis, as has been shown in

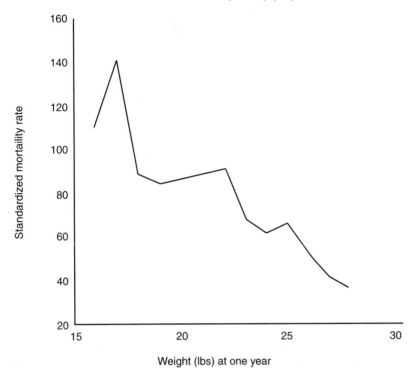

Figure 2.2 The standardized mortality rate from coronary heart disease in a British population by weight at one year of age (Barker 1995).

undernourished laboratory animals (Langley and Jackson 1994). Increased HPA axis activity (as measured by high cortisol levels) has been correlated with both thinness at birth and symptoms of the insulin resistance syndrome such as reduced glucose tolerance, increased blood pressure, and abnormal lipid metabolism (Phillips *et al.* 1998).

Despite the growing body of evidence supporting the Barker hypothesis, there remain many critics who point to both methodological issues involving the use of longitudinal data and misinterpretations of data as problematic (Lucas *et al.* 1999). Some early studies were marred by selection bias, in that mortality-related data were unavailable for a good portion of the birth cohort studied (Barker *et al.* 1989). This problem appears to have been resolved in more recent studies, however, such as a Swedish study in which researchers successfully tracked 97 per cent of the cohort, finding significant correlations between birth weight and mortality from heart disease (Leon *et al.* 1998). Other common criticisms include failure to control for gestational age (Krämer 2000), statistical inaccuracies (Lucas *et al.* 1999), and the use of indirect measures (birth weight, ponderal index, etc.) to assess fetal nutrition (Paneth and Susser 1995).

Perhaps the greatest problem, however, is the tendency to assign causal relationships when the evidence, thus far, is only correlational. Two potential

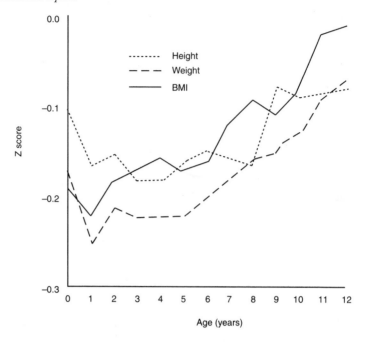

Figure 2.3 Catch-up growth in height, weight, and BMI for Finnish boys who later developed coronary heart disease (Eriksson *et al.* 2001).

confounding variables, socioeconomic status and genetics, recur in criticisms of the fetal origins hypothesis. Critics argue that low socioeconomic status may underlie both low birth weights and a tendency towards particular behavioural risk factors (i.e. diet, exercise, smoking) conducive towards heart disease later in life (Krämer 2000). A number of studies, however, have attempted to parcel out the effects of socioeconomic status and have found the correlations still hold (Leon *et al.* 1998; Eriksson *et al.* 1999).

Genetics may also play a role that is unaccounted for by the Barker hypothesis. For instance, a mother may have a genetic tendency towards high cholesterol that may then be passed along to her offspring irrespective of fetal environment and nutritional status. In this vein, Elbein (1997) notes the heritability of insulin resistance, citing genes responsible for reduced insulin secretion and their putative loci. Similarly, a recent study showed that mothers who give birth to small babies or premature babies, or who suffer from pre-eclampsia are themselves more likely to suffer from coronary heart disease. Perhaps, the authors hypothesize, these mothers and infants share genetic predispositions towards heart disease that are unrelated to the infants' fetal environments (Smith *et al.* 2001).

The Barker hypothesis clearly owes its existence to the availability of longitudinal data allowing prenatal and perinatal conditions to be linked to later disease outcomes. It also rests theoretically on the concept of trade-offs in energy allocation. While the disease outcomes may be viewed as pathological, the trade-offs themselves

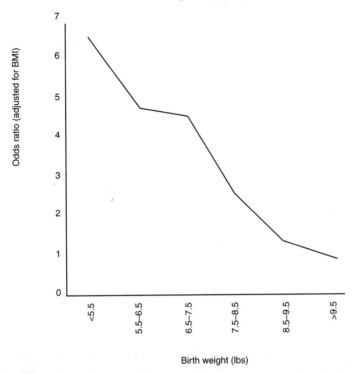

Figure 2.4 Odds ratio (adjusted for BMI) for the development of non-insulin-dependent diabetes mellitus in a British population by birth weight (Barker 1995).

may be functional in a broader life history perspective. That is, it is possible to interpret the patterns of fetal growth as functional adjustments to undernutrition, making the best of a bad situation. Restricting fetal growth may have costs in terms of health risks later in life, but those costs may be outweighed by the benefit of avoiding fetal death or major abnormalities of brain development.

Trade-offs in immune function

Immune system development provides another example of physiological trade-offs with life historical consequences. The two classical branches of the mammalian immune system – cell-mediated immunity and humoral immunity – are often represented as functionally different modes of response to different types of infection. Cell-mediated immunity, in this simplified view, represents a response to intracellular (viral or parasitic) infection. It involves the activation of natural killer cells, phagocytotic macrophages, and complement-mediated cytotoxicity. Humoral immunity, in contrast, represents a response to extracellular (bacterial) infection involving mast cell and eosinophil activation and stimulation of B-cell production of M, E, G1 and G4 immunoglobulins.

In the past decade it has become clear that alternative pathways of helper T-cell differentiation regulate these modes of immune function. Briefly, two distinct populations of helper T-cells can be identified on the basis of the cytokine signals they produce in response to antigen stimulation (Figure 2.5). The type 1 helper T-cells (Th1) respond by producing gamma interferon, tumor necrosis factor beta, and interleukin 2. In combination, these cytokines stimulate phagocytotic activity of the macrophages and the associated delayed-type hypersensitivity (DTH) response, cytotoxic activity of natural killer cells, and the complement-activating secretion of immunoglobulin G2a by B-cells. Type 2 helper T-cells, on the other hand, secrete interleukins 4, 5, 6, and 10, among others. These signals serve to stimulate eosinophil activity, mast cell activity and the associated immediate hypersensitivity response, and B-cell production of G1, G4, E and M class immunoglobulins. The Th1 response dominates during the activation of classical cell-mediated immunity while the Th2 response dominates during activation of classical humoral immunity (Mosmann and Coffman 1989).

More recently it has become apparent that these two T-cell populations are derived from the same pool of less differentiated precursor cells (Swain *et al.* 1991; Romagnani 1994). Signals associated with initial antigen presentation determine subsequent differentiation of the Th1 or Th2 response on the part of stimulated T-cells. At the same time, the stimulation of either Th response suppresses the other. Gamma interferon production as part of the Th1 response, for example, suppresses development of Th2 cells, while interleukin-4 production as part of the Th2 response suppresses development of Th1 cells. Other cytokines involved in each response suppress down-stream components of the alternative response as well. For example, interleukin-10 not only stimulates B-cell production of the immunoglobulins typical of the Th2 response, it also suppresses macrophage activity typical of the Th1 response. Similarly, gamma interferon not only stimulates macrophage and natural killer cell activity as part of the Th1 response, it suppresses B-cell production of IgE typical of the Th2 response.

These two patterns of immune response thus represent alternative and competing trajectories of helper T-cell development. The existence of these alternative trajectories can be viewed as an adaptive feature of the immune system allowing optimization of response to correspond to different infectious challenges. The more acute and exocytotic responses that are appropriate to bacterial infections, i.e. those typical of the Th2 response, are less well suited to the more chronic and endocytotic challenges of viral and parasitic infections, and vice versa. Mounting either immune response is metabolically expensive however, requiring irreversible allocation of both energy and leukocyte resources. Constraints on the allocation of these resources lead to a trade-off between these two responses. They cannot be simultaneously maximized. Rather, natural selection appears to have designed mechanisms to optimize the balance between them.

Appreciating the nature of the Th1/Th2 trade-off can be important to efforts to artificially manipulate immune responses. Vitamin A supplementation, for example, has been observed to dramatically reduce childhood morbidity and mortality in many developing countries, particularly morbidity and mortality

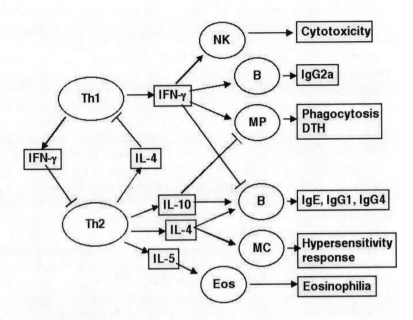

Figure 2.5 Effects of Th1 and Th2 cells on immune function. NK, natural killer cells; B, B
cells; MP, macrophage cells; MC, mast cells; Eos, eosinophils. Adapted from Long
and Santos (1999).

associated with diarrheal disease. This result is likely due to the effect of vitamin
A in stimulating a strong Th2 response (Long and Santos 1999). The Th2 response
is particularly effective in combating bacterial agents of diarrhoeal disease, such
as *Vibrio cholera*. However, as Long and Santos point out, diarrhoeal disease can
also result from infection by intracellular agents such as *Salmonella*. Not only would
the Th2 response promoted by vitamin A supplementation be less effective in
addressing such an infection, it would also lead to a suppression of the more effective
Th1 response on the part of the host, thus actually exacerbating the situation.
Preliminary results indeed suggest just such a consequence of vitamin A supplementation in areas of Mexico with a high prevalence of *Salmonella* infection (Long,
personal communication). Other micronutrients can have similar effects in
stimulating one or the other pattern of Th response. Vitamin D3 appears to
promote a Th2 response and to suppress a Th1 response in a fashion similar to
vitamin A. Vitamins B6, C, and E appear, on the other hand, to promote Th1
responses and to suppress Th2 responses (Long and Santos 1999).

Although the trade-off between Th1 and Th2 immune responses can be
relatively short-lived, coincident with episodes of acute infection, there has also
been speculation that longer-term developmental biases in immune response can
be conditioned during the maturation of the immune system in childhood
(Romagnani 1992; Holt 1995). Biasing subsequent patterns of immune response

in this way could be functional if the pattern of immune challenges in childhood is a good predictor of the pattern of immune challenges later in life. Paradoxical consequences can ensue, however, especially when the exposure of infants and children to infectious agents in the environment is controlled or manipulated. In particular, it has been suggested that a reduction in exposure to parasitic disease in developed countries may have led to reduced stimulation of Th1 lymphocyte development during critical periods of immune system maturation in childhood (Martinez 1994; Strachan *et al.* 1997; Cookson and Moffatt 1997). It has been noted by Hurtado and others (Hurtado *et al.* 1997, 1999) that indigenous populations with high burdens of parasitic infection have extremely low prevalences of asthma and other allergic hypersensitivities. Other studies suggest that exposure to viral infections in early childhood can also have a lasting effect on the alternative trajectories of immune system development. A German study, for example, found that children who are sent to group day care facilities in their first year suffer higher rates of viral infection during that time but have lower rates of asthma and allergic symptoms later in childhood (Krämer *et al.* 1999).

Clearly, the trade-offs implicit in the physiological mechanisms of immune function have explicit consequences for human disease. These trade-offs are intrinsically life-historical because the developmental history of the organism influences its future responses. They also present a clear example of mechanisms designed by natural selection to optimize the allocation of physiological resources over the life cycle of the organism.

Trajectories of reproductive function and their disease consequences

Reproductive trade-offs lie at the very heart of evolutionary life history theory. Trade-offs can be identified between reproduction and survival, between reproduction now and reproduction later, between investing in offspring already conceived or born (parental effort) and investing in additional conceptions or offspring (mating effort), between investing in one's own offspring (direct effort) or the offspring of others (indirect effort), between offspring quality and offspring quantity, and numerous other pairs of competing alternatives. Many of these trade-offs are mediated at more than one level of biology, including genetic, physiological, and behavioural levels. Gonadal function provides a particularly illustrative example, at the physiological level, of alternative developmental trajectories that have consequences for human disease.

The gonadal function of males and females can be indexed by the secretory profiles of the major gonadal sex steroids, testosterone for males, oestradiol and progesterone for females. There is now ample evidence that levels of steroid secretion vary within individuals, between individuals, and between populations in consistent patterns (Ellison 2001). One pattern reveals a positive correlation between metabolic energy availability and gonadal function. Another pattern associates variation in gonadal function with age. There are interesting differences in both of these patterns between males and females, but within each sex the patterns are very consistent across a broad range of cultural and ecological settings and genetic backgrounds. As such they can be considered basic features of human biology.

In combination, however, these two patterns of variation produce variation in trajectories of gonadal function over the life cycle. In populations where energy availability is high, the trajectory of gonadal function by age is particularly high, while the trajectory is low when energy availability is low (Ellison *et al.* 1993). In women, conditions of high energy availability are associated with earlier age at menarche, higher levels of gonadal steroids during middle adulthood, and a steeper decline in steroid levels in the approach to menopause compared with conditions of energy scarcity (Ellison 1996a, 1996b; Figure 2.6). In males conditions of high energy availability are associated with an earlier growth spurt, higher testosterone levels in young adulthood, and a steeper decline in testosterone with advancing age compared with conditions of energy scarcity (Ellison *et al.* 1998; Figure 2.7). Although much of the evidence for these developmental trajectories is cross-sectional, an increasing body of longitudinal data supports the same patterns.

It has been suggested that these trajectories represent developmental optimization of gonadal function to ecological circumstances (Ellison 1990, 1996a, 1996b, 1998, 2001; Bribiescas 1996, 2001). In the case of females, energy abundance shifts the optimal allocation of energy toward higher fecundity at all ages, a shift supported by higher levels of gonadal function. In males, energy abundance shifts the optimal allocation of energy toward increased mating effort particularly in young adulthood, a shift supported by higher testosterone levels and the associated increases in somatic and behavioral components of male reproductive effort. Conditions of restricted energy availability shift optimal energy allocation in both sexes toward survival and away from reproductive effort.

This view implies that energy allocation to reproduction does have consequences for survival. The immediate consequences of reproduction–survival trade-offs may not be apparent in any given population since a positive correlation may be expected between the absolute level of energy allocation to each of these categories. That is, conditions of energy abundance may lead to an increase in the proportional allocation of energy to reproduction, but they may also lead to an absolute increase in energy allocation to survival. However, other unavoidable negative health consequences of increased levels of gonadal function can be identified at later points in the life cycle.

One negative health consequence of high trajectories of gonadal function results from the cumulative effects of high levels of lifetime steroid exposure (Ellison 1999). Steroids are potent stimulators of mitosis in many of their target tissues. In many instances, such as the endometrial lining of the uterus, the lobulo-alveolar ducts of the breast, or the secretory epithelium of the prostate gland, this mitotic stimulation serves to maintain and renew highly active secretory tissues. An unavoidable consequence of this mitotic stimulation, however, is a greater risk of carcinogenic mutation and a greater simulation of the growth of incipient tumors. In both males and females, reproductive cancer incidence is positively associated with endogenous steroid exposure. This is particularly well-studied in the case of

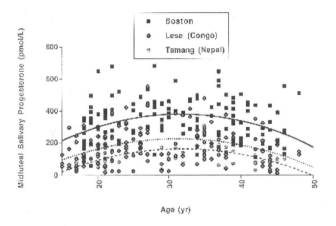

Figure 2.6 Midluteal salivary progesterone levels by age in women from three populations: Boston women, Lese horticulturalists from Congo, and Tamang agropastoralists from Nepal (Ellison *et al.* 1993).

breast cancer in women, where virtually every factor that modulates life-time gonadal steroid exposure has been linked to breast cancer risk. Jasienska and Thune (2001) have recently shown that population variation in breast cancer risk is strongly positively associated with population variation in gonadal steroid exposure during the prime reproductive years (Figure 2.8). Similarly, prostate cancer risk in men is highest in populations with high trajectories of gonadal function.

A second category of negative health consequences of high trajectories of gonadal function is more speculative (Ellison 1996a). Here we must take note of the fact that the levels of gonadal function of men and women vary less in old age than they do in young adulthood. Thus higher levels of gonadal steroids early in life are associated with steeper trajectories of decline with advancing age. A number of negative health consequences are associated with declining steroid levels in both males and females: loss of lean body mass and increasing central adiposity, loss of bone mineral density, and increasing risk of heart disease and senile dementia are among the most notorious. If other aspects of adult physiology adjust to complement an individual's adult level of gonadal steroid production, receptor densities in non-reproductive tissues for example, then the different absolute changes associated with aging in individuals with different steroid trajectories may have different consequences. Individuals whose bone physiology is set to correspond to high steroid levels may be more at risk of loss of bone mineral density with advancing age. Similar differential effects may affect the maintenance of lean body mass or the loss of hippocampal neurons in individuals following different steroid trajectories. In all these cases, the negative consequences of a greater absolute fall in steroid levels in old age would be associated with the positive fitness consequences of higher steroid trajectories early in life: faster growth, larger size,

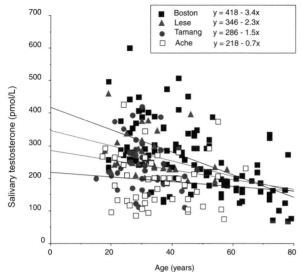

Figure 2.7 Morning salivary testosterone levels by age in men from four populations: Boston men, Lese men, Tamang men, and Ache hunter-gatherers from Paraguay (Ellison *et al.* 1998).

higher fecundity and reproductive effort. As in the case of reproductive cancer risk, these positive fitness benefits may not be fully realized in modern society. Nevertheless these links between earlier and later effects of variation in steroid trajectories represent clear examples of life history trade-offs.

Summary

A life history perspective on human disease has enormous heuristic value, both empirically and theoretically. It directs attention toward the long-term consequences of early conditions and developmental trajectories, and toward the trade-offs in resource allocation and outcomes that have been sculpted by natural selection. A life history perspective is difficult to operationalize, since it requires longitudinal data that are often difficult or impossible to reconstruct retrospectively and difficult and expensive to develop prospectively. Yet there is enormous potential to be realized which should motivate us to confront these challenges. By better discriminating the genetic, developmental, and acute sources of disease we may be better able to develop effective strategies of prevention and treatment. Should the Barker hypothesis be validated, for example, it may prove a more effective public health strategy to combat coronary heart disease through improved prenatal diet and health care than through drastic surgical intervention later in life.

However, the life history perspective carries a trade-off for us as human biologists as well. While it illuminates many aspects of human disease experience it simultaneously undermines the very dichotomy between health and disease on which so much of western biomedicine is founded. As long as disease can

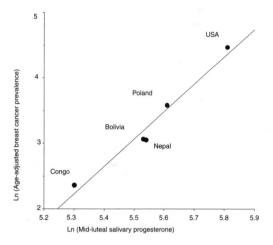

Figure 2.8 The natural log of age-adjusted breast cancer prevalence plotted against the natural log of mid-luteal salivary progesterone in pmol/L (Jasienska and Thune 2001).

comfortably be defined as a pathological state, then health can be defined as its absence and the practice of medicine and public health can be directed toward the avoidance and correction of pathological states. The life history perspective, however, suggests that trade-offs must often be faced not between health and disease, but between alternative 'pathological' outcomes. The debate surrounding hormonal replacement therapy in post-menopausal women represents the difficulty that such trade-offs can present to the traditional, dichotomous concept of health and disease. Similarly, it may be found that the developmental perspective on disease that the life history approach encourages embodies equivalent paradoxes. The developmental trajectories that lead to positive health outcomes early in life – strength, vigor, high fecundity – may have associated negative consequences later in life – reproductive cancers, osteoporosis, Alzheimer's disease. There may be no path that leads unambiguously away from disease and toward health. Rather, we are left with an appreciation for the ineluctable trade-offs that shape human life histories and underlie human evolution.

References

Bakketeig, L.S. and Hoffman, H.J. (1981) 'Epidemiology of preterm birth: results from a longitudinal study of births in Norway', in M.G. Elder and C.H. Hendricks (eds) *Preterm Labor*, London: Butterworth.

Barker, D.J. (1994) *Mothers, babies and disease in later life*, London: BMJ Publishing Group.

Barker, D.J. (1995) 'Fetal origins of coronary heart disease', *British Medical Journal*, 311: 171–4.

Barker, D.J., Winter, P.D., Osmond, C., Margetts, B. and Simmonds, S.J. (1989) 'Weight in infancy and death from ischaemic heart disease', *Lancet*, 2: 577–80.

Bribiescas, R.G. (1996) 'Salivary testosterone levels among Aché hunter/gatherer men and a functional interpretation of population variation in testosterone among adult males', *Human Nature*, 7: 163–88.

Bribiescas, R.G. (2001) 'Reproductive physiology of the human male: an evolutionary and life history perspective', in P.T. Ellison (ed.) *Reproductive Ecology and Human Evolution*, Cambridge, MA: Harvard University Press.

Cookson, W.O.C.M. and Moffatt, M.F. (1997) 'Asthma: an epidemic in the absence of infection?', *Science*, 275: 41–2.

dos Santos Silva, I., De Stavola, B.L., Mann, V., Kuh, D., Hardy, R. and Wadsworth, M.E. (2002) 'Prenatal factors, childhood growth and age at menarche', *International Journal of Epidemiology*, 31: 405–12.

Elbein, S.C. (1997) 'The genetics of human non-insulin dependent (type 2) Diabetes Mellitus', *Journal of Nutrition*, 127: 1891S–6S.

Ellison, P.T. (1990) 'Human ovarian function and reproductive ecology: new hypotheses', *American Anthropologist*, 92: 933–52.

Ellison, P.T. (1996a) 'Developmental influences on adult ovarian function', *American Journal of Human Biology*, 8: 725–34.

Ellison, P.T. (1996b) 'Age and developmental effects on adult ovarian function', in L. Rosetta, and N.C.G. Mascie-Taylor (eds) *Variability in Human Fertility: A Biological Anthropological Approach*, Cambridge, UK: Cambridge University Press, pp. 69–90.

Ellison, P.T. (1998) 'Age change in reproductive function' in S.J. Ulijaszek, F.E. Johnston and M.A. Preece (eds) *Cambridge Encyclopedia of Human Growth and Development*, Cambridge, UK: Cambridge University Press, pp. 422–423.

Ellison, P.T. (1999) 'Reproductive ecology and reproductive cancers', in C. Panter-Brick and C. Worthman (eds) *Hormones and Human Health*, Cambridge, UK: Cambridge University Press, pp. 184–209.

Ellison, P.T. (2001) *On Fertile Ground*, Cambridge, MA: Harvard University Press.

Ellison, P.T., Panter-Brick, C., Lipson, S.F. and O'Rourke, M.T. (1993) 'The ecological context of human ovarian function', *Human Reproduction*, 8: 2248–58.

Ellison, P.T., Lipson, S.F., Bribiescas, R.G., Bentley, G.R., Campbell, C. and Panter-Brick, C. (1998) 'Inter- and intra-population variation in the pattern of male testosterone by age', *American Journal of Physical Anthropology*, Suppl. 26: 80.

Eriksson, J.G., Forsén, T., Tuomilehto, J., Osmond C. and Barker, D.J.P. (1999) 'Catch-up growth in childhood and death from coronary heart disease: longitudinal study', *British Medical Journal*, 318: 427–31.

Eriksson, J.G. Forsén, T., Tuomilehto, J., Osmond, C. and Barker, D.J.P. (2001) 'Early growth and coronary heart disease in later life: longitudinal study', *British Medical Journal*, 322: 949–53.

Fall, C.H., Vijayakumar, M., Barker, D.J., Osmond C. and Duggleby, S. (1995) 'Weight in infancy and prevalence of coronary heart disease in adult life', *British Medical Journal*, 310: 423–7.

Gadgil, M. and Bossert, W.H. (1970) 'Life historical consequences of natural selection', *American Naturalist*, 104: 1–24.

Godfrey, K.M. and Barker, D.J. (2000) 'Fetal nutrition and adult disease', *American Journal of Clinical Nutrition*, 5: 1344S–52S.

Godfrey, K.M., Hales, C.N., Osmond, C., Barker, D.J. and Taylor, K.P. (1996) 'Relation of cord plasma concentrations of proinsulin, 32–33 split proinsulin, insulin, and C-peptide to placental weight and the baby's size and proportions at birth', *Early Human Development*, 46: 129–40.

Hamilton, W.D. (1966) 'The moulding of senescence by natural selection', *Journal of Theoretical Biology*, 12: 12–45.

Holt, P.G. (1995) 'Postnatal maturation of immune competence during infancy and childhood', *Pediatric Allergy and Immunology*, 6: 59–70.

Hurtado, A.M., Hill, K., Hurtado, I.A. and Rodriguez, S. (1997) 'The evolutionary context of chronic allergic conditions', *Human Nature*, 8: 51–75.

Hurtado, A.M., Hurtado, I.A., Sapiien, R. and Hill, K. (1999) 'The evolutionary ecology of childhood asthma', in W. Trevathan, E.O. Smith. and J. McKenna (eds) *Evolutionary Medicine*, New York: Oxford University Press.

Jasienska, G. and Thune, I. (2001) 'Lifestyle, hormones, and risk of breast cancer', *British Medical Journal*, 322: 586–7.

Krämer, M.S. (2000) 'Invited commentary: association between restricted fetal growth and adult chronic disease: Is it causal? Is it important?', *American Journal of Epidemiology*, 152: 605–8.

Krämer, U., Heinrich, J., Wjst, M., and Wichmann, H.-E. (1999) 'Age of entry to day nursery and allergy in later childhood', *Lancet*, 353: 450–4.

Langley, S.C. and Jackson, A.A. (1994) 'Increased systolic blood pressure in adult rats induced by fetal exposure to maternal low protein diets', *Clin Sci*, 86: 217–22.

Leon, D.A., Lithell, H.O., Vagero, D., Koupilova, I., Mohsen, R., Berglud, L., Lithell, U.B. and McKeigue, P.M. (1998) 'Reduced fetal growth rate and increased risk of death from ischaemic heart disease: cohort study of 15,000 Swedish men and women born 1915–1929', *British Medical Journal*, 317: 241–5.

Long, K.Z. and Santos, J.I. (1999) 'Vitamins and the regulation of the immune response', *Pediatr. Infect. Dis. J.*, 18: 283–90.

Lucas, A., Fewtrell, M.S. and Cole, T.J. (1999) 'Fetal origins of adult disease – the hypothesis revisited', *British Medical Journal*, 319: 245–9.

Martinez, F.D. (1994) 'Role of viral infections in the inception of asthma and allergies during childhood: could they be protective?', *Thorax*, 49: 1189–91.

Mosmann, T.R. and Coffman, R.L. (1989) 'Th1 and Th2 cells: different patterns of lymphokine secretion lead to different functional properties', *Annual Review of Immunology*, 7: 145–73.

Paneth, N. and Susser, M. (1995) 'Early origins of coronary heart disease (the "Barker hypothesis")', *British Medical Journal*, 310: 411–12.

Phillips, D.I., Barker, D.J., Fall, C.H., Seckl, J.R., Whorwood, C.B., Wood, P.J. and Walker, B.R. (1998) 'Elevated plasma cortisol concentrations: a link between low birth weight and the insulin resistance syndrome?', *Journal of Clinical Endocrinology and Metabolism*, 83: 757–60.

Romagnani, S. (1992) 'Human Th1 and Th2 subsets: regulation of differentiation and role in protection and immunopathology', *International Archives of Allergy and Immunology*, 98: 279–85.

Romagnani, S. (1994) 'Lymphokine production by human T cells in disease states', *Annual Review of Immunology*, 12: 227–257.

Smith, G.C., Pell, J.P. and Walsh, D. (2001) 'Pregnancy complications and maternal risk of ischaemic heart disease: a retrospective cohort study of 129,290 births', *Lancet*, 357: 2002–6.

Strachan, D.P., Harkins, L.S., Johnston, I.D.A. and Anderson, H.R. (1997) 'Childhood antecedents of allergic sensitization in young British adults', *Journal of Allergy and Clinical Immunology*, 99: 6–12.

Swain, S.L., Bradley, L.M., Croft, M., Tonkonogy, S., Atkins, G., Weinberg, A.D., Duncan, D.D., Hedrick, S.M., Dutton, R.W. and Huston, G. (1991) 'Helper T-cell subsets: phenotype, function and the role of lymphokines in regulating their development', *Immunology Reviews*, 123: 115–44.

Williams, G.C. (1966) 'Pleiotropy, natural selection, and the evolution of senescence', *Evolution*, 11: 398–411.

3 The evolution, transmission and geographic spread of infectious diseases in human populations

Questions and models

Lisa Sattenspiel

Mathematical modeling is one of the newest methods being used to understand how and why infectious diseases evolve and spread across the landscape. The development of this approach has been aided by dramatic improvements in computer power over the last quarter century. Mathematical modeling is a general approach that allows researchers to 'experiment' and test hypotheses about the mechanisms for disease evolution and transmission, without having to use the obviously unethical approaches of real human experimentation.

This chapter will begin by describing some of the major questions being addressed by anthropologists, mathematicians, and biologists interested in the evolution, transmission, and spread of infectious diseases. Following this, the principal modeling techniques used to answer these questions will be described. The chapter will conclude by illustrating the process of infectious disease modeling with a study of the impact of population travel patterns on the spread of a 1984 measles epidemic on the West Indian island of Dominica.

Ecological and evolutionary issues in infectious disease modeling

Population biologists traditionally study three fundamental ecological and evolutionary interactions: competition within and between species, predator–prey interactions, and the interaction between a pathogen and its host species. Mathematical modeling has been an important component in the study of these interactions for a century or more. Both mathematicians and ecologists have developed these models, resulting in a range of work from models focused on elegant but not necessarily practical formulations to those motivated by the need for answers to specific real-world problems. Underlying the best of these is a firm understanding of the essential biology of the ecological interaction considered. Consequently, before addressing the types of questions and models used in infectious disease modeling it is helpful to look at the nature of the host–pathogen interaction.

Host–pathogen relationships

The essential ecological relationships in infectious diseases involve the interaction between a pathogen, its host(s), and the environment. By definition, a pathogen is an organism that causes disease, thereby having a negative impact on its host. A number of different strategies, called modes of transmission, are used by pathogenic organisms to spread from one host to another. At the same time, host organisms have evolved mechanisms to promote or inhibit transmission of pathogens. Because of this, humans and their pathogens have evolved a wide diversity of strategies in order to ensure survival.

Modes of transmission of human infectious diseases can be either direct, where a pathogen is transmitted directly from one person to another, or indirect, where transmission occurs by means of an intermediate host or agent. Direct modes of transmission include respiratory, fecal–oral, sexual, and congenital transmission, as well as transmission by direct physical contact. Major modes of indirect transmission include water-borne, food-borne, soil-borne, and vector-borne transmission, needle-sharing, and more complex life cycles that involve several different hosts and vectors.

Diseases spread by any of these modes of transmission can be found in all regions of the world. However, the majority of common diseases in temperate regions are transmitted directly from person to person, while many of the most common tropical diseases are spread indirectly, especially through vector-borne transmission. This latter mode of transmission occurs when a non-vertebrate host (most often an arthropod) either mechanically carries a pathogen from one host to another or serves as a secondary host in which the pathogen completes part of its life cycle. Mosquitoes, ticks, fleas, lice, sand flies, midges, and gnats are the most common vectors (Chin 2000), with malaria, yellow fever, dengue fever, trypanosomiasis, and filariasis some of the most common vector-borne diseases.

A number of reasons exist for differences in the predominant types of infectious diseases and transmission modes in different regions of the world, including reasons related both to differences in the natural environment and to the nature of human activities and behaviors. For example, levels of biodiversity influence the availability of potential hosts and pathogens, temperature and humidity affect the rate of reproduction of both hosts and pathogens, world regions vary in their general levels of sanitation and personal hygiene, economic factors influence the quality of health care available, etc.

These factors, both environmental and cultural, lead to a wide diversity of interactions among human hosts, their pathogens, and vectors or alternate hosts of those pathogens. In addition, humans have a much longer generation time than other species in the transmission cycle. This has major consequences for the human host–pathogen interaction. Other species in the cycle, especially the pathogens, tend to evolve new ways to spread from one person to another faster than the human body can evolve means to evade transmission. Because of this, most strategies humans use to deal with infectious diseases are culturally driven. Nonetheless, the essential human host–pathogen interaction is biological, making the study of infectious disease ecology in human populations a fundamentally biocultural endeavor.

Major evolutionary issues in infectious disease research

The rapid evolution of pathogens relative to the generation time of most of their hosts has given rise to seven major directions for studies of the evolution of infectious disease. These include: 1) the mechanisms by which new strains and species of pathogens or reservoirs evolve; 2) the evolution of virulence; 3) the evolution and impact of different modes of transmission; 4) the development and consequences of antibiotic resistance; 5) the origin, relatedness, and routes of transmission of different strains of infectious organisms; 6) the role of pathogens in the origin of sexual behaviour; 7) and the impact of sexual selection on the evolution of pathogens. The first four of these categories will be addressed briefly below, with emphasis on applications, if any, to diseases affecting human populations. Studies addressing the origin and relatedness of different strains of infectious organisms are based primarily upon the methods of genetics and molecular phylogeny, while studies addressing the latter two categories fall more traditionally within the realm of behavioral biology rather than epidemiology. Because they are peripheral to the main focus of this chapter, these three areas will not be discussed further.

Much has been written about the recent apparent increase in the emergence of new infectious diseases or re-emergence of those once thought to have been controlled (cf. Garrett 1994; Levins *et al.* 1994; Wilson *et al.* 1994; Wills 1996; World Health Organization 1996). A number of factors are implicated in this rise, including, for example, increased rates of worldwide travel and shipping, destruction of natural environments that support the nonhuman hosts of pathogens, climate change, and inadequate access to or utilization of health care resources. This situation has magnified the need to understand the mechanisms by which new strains and species of pathogens evolve. However, this is an area that has received relatively little attention from mathematical epidemiologists, who have focused more on ecological issues related to infectious disease transmission. Some of the earliest models that address the evolution of new strains of pathogens were developed by Anderson and May (1982) and May (1983). Much of the research on this topic is theoretical and does not focus on particular pathogens, although models have been developed for the evolution of different influenza strains (e.g. Pease 1987; Lin *et al.* 1999) and for the origin of HIV (Burr *et al.* 2001).

The evolution of virulence is the most frequent question addressed in evolutionary models of infectious diseases and is one that has often been considered in relation to human infectious diseases. Early researchers on this question assumed that since a virulent pathogen harms the host organism, it must have only recently begun to exploit the host and that over time the host and pathogen would co-evolve so that the pathogen was less virulent and/or the host was more resistant (Dubos 1965; Burnet and White 1972; May and Anderson 1983). This view clearly holds for many pathogens, but there are numerous examples of highly virulent pathogens with a long history of interaction with a given species (Levin 1996). For example, cholera is almost certainly an ancient human disease and has caused repeated and severe worldwide epidemics for at least the last 165 years (Barua 1992). By the early 1980s evolutionary biologists began to recognize that evolution of a pathogen to a more benign form was not a necessary consequence of host–pathogen co-evolution, but that it was possible for selection to favor the maintenance of a more virulent form. These ideas were synthesized and formalized by Ewald (1994, 1996, 1998), who hypothesized that the evolution and maintenance of highly virulent strains would be especially likely when the biology of the parasite is such that transmission could occur even when a host was immobilized. Ewald argued that this explains why cholera remains virulent, because, for example, it can be spread when a sick person's attendant takes dirty sheets to a river for washing. Drawing on the discussions of the factors that would lead to evolution of virulent pathogens, Bull (1994) proposed a set of generalities about virulence, including that there is often a trade-off between the benefits of increased transmission and the costs of decreased virulence, that higher rates of transmission between hosts generally select for greater virulence, and that new pathogens introduced into a species may lead to highly virulent infections.

(handwritten margin note: makes no sense)

Because the evolution of more or less virulent pathogens is often associated with mode of transmission, a significant amount of research on the evolution of infectious diseases has focused on mode of transmission. The most common questions that are addressed include the advantages and disadvantages of horizontal vs. vertical transmission and the implications of sexual vs. nonsexual transmission.

In common with models for the evolution of new strains of pathogens, models for the evolution of virulence and for the influence of modes of transmission on the evolution of pathogens tend to be theoretical in orientation and focused on either nonspecific or nonhuman diseases. Examples of models that specifically address evolutionary issues in human diseases include Kiszewski and Spielman (1994), Lipsitch and Nowak (1995), and Anderson and May (1996).

Antibiotic resistance is a problem of growing concern, but the attention paid to this question by mathematical modelers is limited. However, unlike the other evolutionary issues discussed, the immediate public health implications of antibiotic resistance have guaranteed that the majority of models that have been developed focus on human diseases. Diseases with growing problems as a result of the evolution of antibiotic resistant strains that have benefited from mathematical modeling include drug resistant tuberculosis (e.g. Blower and Gerberding 1998; Lipsitch and Levin 1998), influenza (Stilianakis *et al.* 1998), HSV-2 (Blower *et al.* 1998), cancer (Panetta 1997), and *E. coli* infection (Levin *et al.* 1997; Stewart *et al.* 1998).

Selected ecological questions addressed in infectious disease modeling

Evolutionary infectious disease models are a vibrant area of research in theoretical ecology, but their applications to specific human diseases have been limited. This is also the case for ecological models, except for diseases that have captured human imagination, usually because they are newly emergent or recently re-emergent.

Yet there are many factors in the natural and cultural environment that influence the distribution, abundance, and transmission of infectious organisms, generating many interesting questions that can be addressed with mathematical modeling techniques. Because of the paucity of mathematical models that deal directly with these questions, much of the succeeding discussion in this section will focus on introducing some of the most interesting ecological issues, with reference to relevant models where appropriate. Most of these topics can be subsumed under three categories: 1) the impact of directly human-induced ecological changes (e.g. deforestation, agriculture, or water control) on host–pathogen interactions, 2) the impact of natural ecological changes (e.g. climate change or natural disasters) on host–pathogen interactions, and 3) the relationship between human behaviors and disease transmission.

Ecological changes induced by human activities have been implicated in the alteration of transmission patterns of several important diseases. For example, the spread of human settlements into unpopulated jungles and agricultural activities in these areas has led to increased contact with forest-dwelling mosquitoes that carry a sylvan form of yellow fever. Bands of monkeys (the natural host for yellow fever carrying mosquitoes) enter the fields in search of easily accessible foods, further increasing the risk of transmission of the disease to humans (Beaver and Jung 1985).

The incidence of malaria has increased in recent decades in many parts of the world, but the reasons have varied in different places. For example, malaria incidence is higher in communities bordering cultivated swamps in Uganda (Lindblade *et al.* 2000), and in general, over the last 50 years the incidence of the disease has been rising throughout the African highlands. Lindsay and Martens (1998) traced the rise in this region to agroforestry development and scarce health resources. They also used a mathematical model to try to identify epidemic-prone regions and determine alterations in their risk for epidemics as a consequence of global climate change. Cultivation in the Sahel region, on the other hand, destroyed larval sites for the indigenous malaria vector and led to extinction of the disease in that region (Mouchet *et al.* 1998).

Large-scale water control systems have been constructed in many parts of the world as a consequence of the development of agriculture. The resultant changes in the natural environment have been implicated in the increase in transmission of many water-borne pathogens. Dam construction on the River Senegal changed the hydrological characteristics of Lake Guiers, which prevented cultivation of traditional crops and caused a shift to irrigated crops. This further changed the water quality of the lake and increased the frequency of suitable habitats for the snail hosts of schistosomiasis, accelerating the spread of the disease into the neighboring human population (Cogels *et al.* 1997). Near Lake Volta in Ghana schistosomiasis not only increased in frequency, but also shifted in predominant type from the intestinal to the urinary form of the disease. This change in type was a direct consequence of changes in the frequency of different snail hosts in the new environment (Bradley 1998). This example illustrates the complexity of natural ecosystems and the potential unexpected impacts on the entire system when there are changes in the structure of parts of the system.

Mathematical models are particularly useful in the face of such complexity, because they allow a researcher to isolate particular parameters, vary them across a range of values, and assess the importance of those parameters to the overall operation of the system. Schistosomiasis has been the focus of several modeling efforts. Habbema *et al.* (1996) and de Vlas *et al.* (1996) developed a microsimulation model that incorporates the dynamics of the human, snail, and worm populations, the development of immunity in the human host, and the presence of variation in the average per person worm burden, while Hammad *et al.* (1996) developed a neural network model of schistosomiasis transmission. Both of these models were used to predict the outcomes of potential control strategies.

Natural ecological changes may also affect the rates of transmission of infectious diseases. Natural disasters such as earthquakes can destroy sanitation facilities, increasing exposure to fecally transmitted organisms. Natural disasters can also destroy not only human domiciles, but also the habitat of natural hosts of diseases or disease vectors, bringing these organisms into closer contact with humans and resulting in serious epidemics. For example, Tanguy *et al.* (1998) reviewed the literature on deaths as a result of volcanic action since 1783 and found that 30.3 per cent of all recorded deaths were due to either post-eruption famine or infectious disease.

In addition to natural disasters, many scientists believe that temperatures and humidity levels are rising worldwide and that severe weather is becoming more frequent and intense (Epstein 1999). The increases in temperature and humidity generate increased amounts of rainfall, which can lead to more frequent and serious flooding. This then increases the risk of transmission of disease organisms carried in the water and can also overwhelm sanitation systems, leading to further environmental contamination. Global warming and related environmental changes may also promote the growth of algae and associated plankton blooms, which may significantly affect the distribution of cholera, one of the most common and feared water-borne diseases (Colwell and Huq 1994; Mata 1994; Colwell 1996). A number of researchers have also pointed to the influence rising temperatures may have on mosquito and other vector dynamics, extending the natural range of these species and causing expansion of the ranges of the diseases they carry (e.g. Martens *et al.* 1997; Lindblade *et al.* 1999; Reiter 2001). Because a primary goal of mathematical models is to explore questions relating to the impact of changes in individual parameters, such as temperature, the interrelationship between global warming and the spread of malaria and other vector-borne diseases has been an active area of mathematical modeling research (see, for example, Lindsay and Birley 1996; Massad and Forattini 1998; Yang and Ferreira 2000).

The role of specific human behaviors is a third general area in studies of human disease ecology. The list of possible human behaviors affecting infectious disease transmission is endless, so only selected examples will be included in the following and the focus will be on studies in which modeling was a component of the analysis.

Individuals infected with many infectious diseases, especially vector-borne diseases, vary throughout the day in their capacity to pass on the infection to vectors or hosts. For example, the microfilariae of one form of Bancroftian filariasis circulate in the blood during the night, especially between 10 p.m. and 2 a.m., and are present in only minimal numbers at other times (Chin 2000). Thus, this disease is transmitted preferentially to, and by, nocturnally feeding mosquitoes and the activities of the human, pathogen, and mosquito vector are strongly influenced by the dynamics of transmission. Similar patterns are also seen with malaria in many parts of the world. Knowing this, Nakazawa *et al.* (1998) explored the relationships between location of evening activities and rate of malaria transmission in the Solomon Islands. They found no significant association between timing of activities and disease transmission, but did observe that the type of clothes worn was important. They then used a mathematical model to explore the utility of dress practices in controlling the disease.

The patterns of water contact during daily household activities can have a strong effect on transmission of water-borne diseases. For example, Watts (1987) linked differences in water use between males and females and across regions to differences in the risk of transmission of dracunculiasis (guinea worm). Friedman *et al.* (2001) explored the importance of human water contact patterns for the transmission of endemic schistosomiasis in a village in Brazil.

The nature of housing can also have a strong impact on the transmission of some infectious diseases. Chagas' disease, a growing problem in Central and South America, is transmitted to humans through the bite of a bug, *Triatoma infestans*, that lives in houses and bites while residents are sleeping. Cecere *et al.* (1998) associated the density of this vector within a house to the type of thatch used in the roof, the age of the house, the degree of cracking of indoor walls, the presence of chickens and domesticated dogs within the house, and the use of domestic insecticides. They have used mathematical models to explore the efficacy of different control strategies, including both alterations in human behaviors and chemical control methods (Cecere *et al.* 1998; Gurtler 1999; Cohen and Gurtler 2001). Coimbra (1988) also studied the importance of human behaviors and activities on the incidence of Chagas' disease. He found that the features identified by Cecere and colleagues were not sufficient to explain disease prevalence in all locations. Many small lowland populations in South America have little trouble with the disease, in spite of housing features that are significant risk factors in other regions. Coimbra suggested that small settlement size and high village mobility probably play a role in the reduced infection rates in the lowland populations, because those characteristics may help to prevent the bug infestations that are a necessary link in the disease cycle.

At a larger scale, increases in global transportation and the social disruptions of wars and other major social events have had and will continue to have marked consequences for infectious disease transmission. Smallman-Raynor and Cliff (1998a, 1998b, 1999, 2001) have used geographically-based models to study the spread of typhoid fever through US military camps during the Spanish–American War, the spread of a cholera epidemic during the Philippines insurrection in 1902–

4, and the propagation of enteric fever, smallpox, and yellow fever during times of both war and peace in Cuba between 1895 and 1898. They found that the spatial pattern of transmission of these diseases was fundamentally altered as a consequence of military activities.

Increasing attention is being paid to the importance of transcontinental transportation on the spread of infectious diseases (Ostroff and Kozarsky 1998). This transportation includes not only travel of humans from place to place, but also transportation of goods on which infectious particles may hitch a ride. The 1991 cholera epidemic in Peru is thought to have started when a ship unloaded ballast water from Southeast Asia into Peruvian coastal waters (although its near simultaneous appearance along a great expanse of the Peruvian coastline has called that conclusion into question (Colwell 1996)). *Cyclospora cayetanensis* imported to the US on Guatemalan raspberries has been linked to several outbreaks of foodborne diarrheal illnesses (Huston and Petri 2001). Epidemic cholera in the Great Rift Valley region of Burundi has been linked to extensive use of lakes and connecting rivers for both transportation of goods and domestic purposes (Birmingham *et al.* 1997), while *Aedes albopictus*, the primary mosquito vector of dengue fever and other arboviruses, has traveled from Asia to both Africa and the United States on used tires and other shipping containers (Reiter 1998; Zeller 1998). Travel within and among countries has been clearly implicated in the temporal and geographic spread of sexually transmitted diseases, including HIV and syphilis (e.g. Wawer *et al.* 1996; Abdullah *et al.* 2000). Several mathematical models have been developed to look at the role of transportation in carrying infectious diseases from place to place. In addition to the model outlined below, examples include the work of Rvachev and Longini (1985), Longini (1988) and Bonabeau *et al.* (1998) on the geographic spread of influenza, a model developed by Murray and Cliff (1975) to study the geographic spread of measles, and the work of Thomas (2000) and Thomas and Smith (2000) on the multiregional spread of sexually transmitted diseases.

Major approaches to modeling infectious disease transmission

Infectious disease transmission has been modeled using a variety of methods, including mathematical modeling, statistical modeling, and computer simulation. Mathematical models generally work from the ground up. They begin with a set of assumptions about the structure of a system, formalize those assumptions into a set of mathematical equations, and then use those equations to try to reproduce patterns observed in the real world. Statistical models tend to work from the top down. They begin with the observed data and try to find mathematical relationships that can explain the patterns observed in the data. Computer simulations are motivated by and often formally related to underlying mathematical models, but the mathematical equations that make up the underlying model are often not made explicit.

The example below illustrates the application of a mathematical model to infectious disease transmission in a modern population. Consequently, the variety of approaches in this area will be the major focus of the remainder of this section. In addition, because computer approaches are becoming increasingly important in infectious disease modeling, some of these approaches will also be discussed. Because of space limitations, statistical approaches will not be discussed further.

Mathematical modeling and infectious disease transmission

A number of different mathematical methods have been used to model the transmission of infectious disease. Choice of a model depends primarily upon the questions being asked and the modeler's experience with different methods. Models can either focus on a population, in which case individuals within a group are generally assumed to be identical to each other, or they can focus on individuals and treat everyone differently. Population-level models are generally easier to analyze mathematically, but the growing use of computers to aid in model analysis has increased interest in individual-based models. Other common characteristics that are taken into account when developing a mathematical model include: 1) whether a single large population is being modeled or whether there are subpopulations that may vary, 2) whether time is considered to be continuous or is discrete (i.e. measured in distinct steps), and 3) whether parameters are assumed to take on fixed values (such a model is called deterministic) or are assumed to vary randomly (such a model is called stochastic).

Most anthropologists who have modeled infectious disease transmission have used one of two types of mathematical approaches: demographic models, especially life tables, and compartmental epidemic models. In life table models infectious diseases are explicitly considered as one factor influencing the mortality schedule of a population. Such models have usually been used to assess the impact infectious diseases may have had on the age distribution and overall growth of a population. Palkovich (1981) used this approach in her studies of North American Arikara populations, as did Thornton *et al.* (1991) in studies of the impact of smallpox on Native American communities and Whitmore (1992) in studies of the population history of colonial Mexico.

Compartmental models divide a population into subgroups on the basis of disease status (e.g. susceptible, infected, recovered, or exposed) and then explicitly consider the factors influencing rates of flow from one disease state to the next. For example, transfer from the susceptible state to the exposed state occurs as a consequence of transmission of the pathogen, which can be affected by suscep-tibility of the uninfected person, infectivity of the pathogen, the probability and/ or timing of contact between a susceptible person and an infectious person, and other factors. Questions to be addressed with the models usually determine what factors are included in the model structure and what factors are ignored. In addition to the example below and other work of the author, anthropological studies incorporating compartmental models include Upham (1986), Ramenofsky (1987), and McGrath (1988).

In recent years there has been a move toward developing composite models that follow population-level spread of a disease across individual-based structures. These models are generally graph- or lattice-based and place individuals at the nodes of the graph. As in traditional compartmental models, individuals are classified on the basis of their disease status. These models are an extreme version of models where a larger population is divided into subgroups; in this case the subgroups consist of single individuals, so that contact is assumed to occur between individuals rather than between subpopulations composed of homogeneous groups of individuals. Generally speaking, transmission may occur between nearest neighbors only. This approach is more realistic than standard compartmental models, but has mostly been limited to theoretical models and has not been used by anthropologists.

An interesting recent variant of lattice-based models incorporates disease spread into models of the 'small-world phenomenon' (Milgram 1967; Pool and Kochen 1978). Small-world models use graph theory to define a network of interactions that includes both highly clustered and highly random associations between individuals. Results from analyses of these models have shown that any two individuals selected at random anywhere in the world can be linked by a very small number of intermediate acquaintances; hence the name 'small-world' model. The original models focused primarily on communications networks; recent applications to infectious disease transmission have been presented in Watts (1999) and Moore and Newman (2000). Kretzschmar and Morris (1996) and Morris and Kretzschmar (1997) have also considered the connection between individual contact networks and the population-level spread of infectious diseases, with emphasis on modeling the spread of HIV across social networks.

Computer-based approaches to modeling infectious disease transmission

One major limitation of most mathematical approaches is that they assume large populations and/or little heterogeneity among individuals or populations. These assumptions *may* be reasonable for the urban populations of the world, but most anthropological studies involve very small populations. Computer-based approaches that depend on modeling individuals within small populations are receiving increasing attention from anthropologists, although their application to infectious disease transmission remains limited.

Many modern-day mathematical modelers use computer-based numerical models extensively in their analyses. However, this approach normally involves computer implementation of a mathematical model that can be analyzed using traditional mathematical techniques. True computer simulation models are different from these numerical models. They involve individual-based stochastic simulation models of a population of individuals and often do not have an explicit underlying mathematical formulation. The two major approaches to true computer simulation used in anthropological research today are microsimulation and agent-based computer modeling, although only microsimulation has been used substantially in infectious disease research.

Individual-based computer simulation models generally assume that each individual in a population possesses a set of measurable attributes and that these attributes vary among individuals. Randomness of events is also a fundamental characteristic of most of these models so that each run of the model is different. Consequently, the models are run repeatedly to allow recognition of underlying patterns in the data.

In a microsimulation model (sometimes called a Monte Carlo model) the attributes of each individual and the rules for interactions among them are first defined. Changes in the attributes occur with predefined probabilities and when the condition for a potential change is met, a random number is generated. If the random number is equal to or lower than the predefined probability, the change occurs; if it is higher, the attribute remains as it was. For example, the average infectious period for influenza is about five days, so a 20 per cent probability of recovery per day can be assumed. The simulation considers the attributes of each individual each day and if a person is infected, a random number is generated. If the generated number is ≤ 0.20, then the infected individual recovers and moves out of the infectious stage; otherwise that individual remains infectious. At each time unit of the simulation all individuals are evaluated and attributes are changed, if appropriate.

One of the earliest examples of microsimulation in infectious disease modeling was presented by Bartlett (1961), who studied causes of the seasonality of measles epidemics. Sophisticated microsimulation models of community structure, including schools, work activities for adults, and many other features of daily life were developed by Ackerman *et al.* (1984) and applied to the spread of several different infectious diseases, including influenza. Fix (1984) used microsimulation to study the demographic and genetic structure of Semai Senoi populations in Malaysia and included infectious diseases as a primary selective factor.

Agent-based models, also known as individual-oriented, distributed artificial intelligence-based, or artificial societies, are becoming increasingly common and valuable in social science research (Gilbert and Conte 1995; Epstein and Axtell 1996; Kohler 2000). The models consist of three basic components: agents, the environment or social space, and rules that govern interactions among agents (Epstein and Axtell 1996). Agents are entities that collect information about the environment and make decisions or take action based on that information (Doran *et al.* 1994; Russell and Norvig 1995; Kohler 2000). The aim of agent-based models is to identify local-scale mechanisms that generate large-scale phenomena, such as social structure or group-level behaviors (Epstein and Axtell 1996).

Epstein and Axtell (1996) provide one of the few examples of the use of an agent-based model in infectious disease research. Their approach is similar to the traditional microsimulations, but each agent has a changeable immune system that varies according to environmental conditions. In a Monte Carlo simulation model, a fixed relationship occurs between a particular level of immunity and the probability of transmission. If the susceptible individual comes into contact with an infectious individual, a random number is generated and compared with the probability of transmission to determine if transmission occurs. In this approach, transmission only occurs if a suitable random number is generated. In Epstein and Axtell's model, if a neighboring agent is infectious, the model checks to see whether the immune system is sufficient to provide immunity, and if not, the disease is transmitted with 100 per cent probability. Transmission throughout a population is facilitated by allowing agents to move to neighboring unoccupied cells, from which they can interact with new neighbors.

Many other anthropological questions have been addressed with agent-based models. Examples include studies of the development of social hierarchies and centralized decision making (Doran *et al.* 1994; Doran and Palmer 1995), simulations of hunters and gatherers in the Scottish Mesolithic (Lake 2000), models of settlement patterns in Mesa Verde (Kohler *et al.* 2000), and models of Anasazi population structure and history (Dean *et al.* 2000). The interesting topics and the breadth of these examples illustrate the far-reaching possibilities this approach provides for addressing questions about the impact of human behavior and biology on the transmission of infectious diseases.

The process of epidemic modeling: the geographic spread of measles in Dominica

The development and analysis of an epidemic model is a cyclical process that includes formulating interesting questions, developing an appropriate model, gathering the data needed to estimate model parameters, running the computer implementation of the model, analyzing the results, devising new questions, modifying the structure of the model, collecting new data, rerunning the model, analyzing the new results, devising new questions, etc. Mathematical modeling is essentially a technique that aids in developing a more structured way to think about a problem and can be used in almost any research project. A study of the spread of measles on the West Indian island of Dominica will be used to illustrate this technique and how it can be used to address anthropological questions. This discussion will focus not only on the models and their results, but also on the nature of the data used in the computer implementations and some of the important issues that have come up in the course of the research.

The Commonwealth of Dominica is a small Caribbean island located between Guadaloupe and Martinique in the Lesser Antilles. It is about 30 miles long and 17 miles wide, with a rugged interior consisting of five volcanic peaks. In 1984 the island experienced a major measles epidemic with over 200 recorded cases in a population of about 80,000. Records of weekly cases of the disease were available from the Ministry of Health for five of the seven health districts. Figure 3.1 shows that the epidemic spread rapidly throughout the island, but reached different districts at different times. Because of the clear temporal separation of cases in association with the geography of the island, this epidemic was ideal to use as the basis for a geographic model of infectious disease transmission.

The base model used in the project combines a standard SIR (susceptible, infectious, recovered) epidemic model with a submodel to handle intercommunity travel of individuals within a region. The model considers a closed population

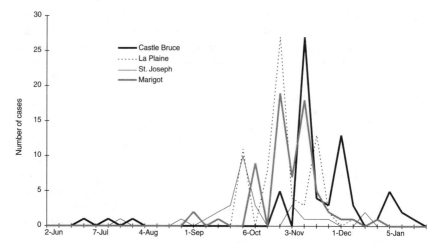

Figure 3.1 Weekly reported cases of measles by health district during the 1984 Dominican epidemic. Only two cases were reported for Grand Bay during 1984 and they both occurred prior to the time plotted. Data from Roseau and Portsmouth were not available.

consisting of the seven Dominican health districts, which are assumed to be constant in size. The residents of each district are distributed into three disease classes: susceptible, infectious, and recovered. Individuals move from the susceptible to infectious stages as a consequence of disease transmission, which is a function of both the probability that a susceptible individual comes into contact with an infectious individual and the probability that transmission occurs given sufficient contact. Transfer out of the infectious class occurs upon recovery from the disease, with recovery conferring permanent immunity to the disease. The mobility submodel is a simple three-parameter model incorporating the rate at which residents of a district travel, the distribution of possible destinations of those travelers, and the return rate from trips. The latter parameter is related to the length of the trip. (See Sattenspiel and Dietz (1995) for details about the specific model structure.)

Analysis of this model requires data to estimate both the epidemiological parameters and the patterns of mobility linking districts. Estimates of the rate at which an infected person recovers from the disease are derived from observations of the amount of time between infection and recovery from measles. This time is well known to be about ten days (Chin 2000), which leads to a recovery rate of 0.10 if it is assumed that a constant proportion (10 per cent) of individuals recover each day. The rate of transmission of measles given contact is much more difficult to estimate and is not directly available from the epidemiological literature. However, because the disease is highly contagious, it is assumed to be 0.30 for all seven health districts. This means that the disease is transmitted 30 per cent of the time a susceptible person comes into contact with an infectious person. The uncertainty in this parameter points out one of the difficulties associated with mathematical modeling – that estimates for model parameters may not be available. In such cases, the appropriate literature (in this case epidemiological) is consulted carefully so that reasonable assumptions can be made about the uncertain parameter. The parameter is then varied systematically in different runs of the computer model so that the effects of its uncertainty can be assessed and controlled for in future studies.

Data for the mobility parameters (probability of taking a trip, distribution of destinations, and length of a trip) were collected in a field study conducted during the winter of 1991. Data were derived from information collected during interviews with over 300 people from all districts on the island. Respondents were asked how often each year they visited each of about 30 towns and sights on the island. Data on the travel patterns of all respondents from a given district were combined to generate average rates of travel from that district and a reasonable estimate of the distribution of destinations for those trips. Because of an oversight at the time of data collection, there was little information about the average length of a trip, but anecdotal evidence and personal observations suggested that it was reasonable to assume that most trips average one day.

One of the difficulties associated with the data collection methods was that most information was collected from adult women, resulting in reasonable estimates for their travel but insufficient data to estimate travel of adult men or children. However, measles is a disease of childhood, so that children's mobility is likely to be of greater importance in spreading the disease than adults' travel. Two sources of information were used to derive estimates of children's mobility. First, several questions were asked about children's travel in the interviews with adult females. Second, interviews were conducted with school officials to determine how often and where children traveled for school-related activities. Dominican children attend school about 200 days a year, or about 55 per cent of their potential travel time. The remaining 45 per cent of their time is spent at home and it is reasonable to assume that they would travel according to the same patterns as their parents during that time. Their overall travel patterns were thus estimated by taking a weighted average of the travel rates derived from data on adults' activities and those derived from school-related travel and children's travel reported by adults, with a 55 per cent weight given to the latter type of travel and a 45 per cent weight given to the adult travel pattern.

Broadly speaking, adults' and children's travel patterns were similar, which is not surprising since 45 per cent of the time children were assumed to follow the adult patterns. However, there were some differences between them. Adults traveled about twice as often as children and traveled predominantly to districts adjacent to their home district. Furthermore, adults from all districts traveled to the island's capital on a regular basis, with over 90 per cent of the respondents doing so at least once a month. Children's predominant destinations were similar to those of adults, but a much higher proportion of their travel involved the northwest and east central districts, primarily because these areas contain the major attractions for school-related field trips.

Simulations of the mathematical model were used to address two major questions: 1) were the observed patterns of the 1984 measles epidemic a reflection of the patterns of travel observed on the island, and 2) what were the effects of differences in patterns of travel of adults and children on overall epidemic patterns? These questions are specific enough to be manageable, but general enough that the answer was not obvious prior to the start of the research. As such, they are representative examples of the kinds of questions that help guide modeling activities.

Once the initial parameter estimates are determined and the main questions are formulated, the stage is set to begin simulations. At the start of the simulations all members of the population are assumed to be at residence in their home health districts. The disease begins with one infected individual in one of the health districts and then is followed as it spreads throughout the districts until it either dies out or reaches an equilibrium state where the numbers of individuals in each category remain constant (or predictably fluctuate). Simulations can be run to explore the consequences of varying any of the model parameters, and although it is possible to vary all of the parameters at once, usually only one parameter at a time is varied to assess the impact of each parameter on disease patterns. This makes it easier to evaluate the impact of interactions among two or more factors.

Ideally, simulations should be run to evaluate the consequences of varying both the initial conditions of a model and the values of each parameter of the model. In reality, this process, although important, is time-consuming and is curtailed in favor of simulations addressing specific questions of interest to the researcher. The effects of population mobility, for example, are influenced by where an epidemic starts, the frequency of travel, where the travelers go during their trips, how long the trips last, etc. Varying all of these factors simultaneously would lead to results that would be difficult or impossible to interpret because it would be hard to determine the extent to which each of the factors operated on the system. To simplify this relatively complex process, the initial simulations of measles in Dominica centered on the effect the location of the starting point of an epidemic had on patterns of disease transmission. Once this effect was understood, other questions were addressed, such as differences in epidemic patterns in models using children's or adults' mobility.

The most important step in mathematical modeling is probably the analysis of the results from the computer simulations, which usually involves identification of patterns in the results that give insights into how different factors influence epidemic outcomes. Figure 3.2 shows the results of two runs of the model, one using adults' mobility patterns and the other using children's mobility patterns. All other parameters were equal in the two models. Several conclusions can be drawn from these results. First, the shape of simulated epidemics is similar for adults' and children's mobility patterns, with well-defined epidemic peaks occurring nearly simultaneously in all districts. The order in which the peaks occur for the seven districts is independent of both the pattern of mobility and the origin of the epidemic (results not shown for the latter). This order appears to follow a geographically-based sequence, with the capital district, Roseau, as the focal point.

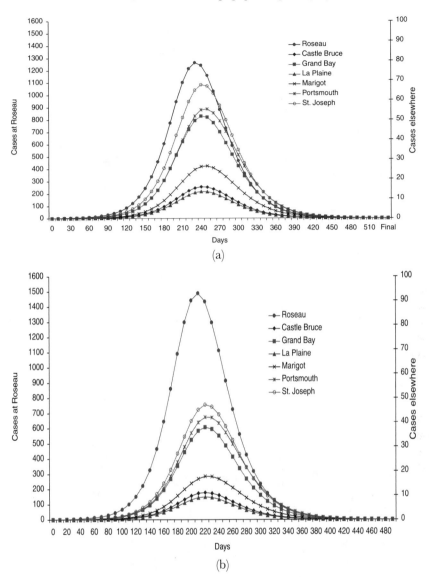

Figure 3.2   Simulated epidemic curves by health district. a) Simulated epidemic using adults'
travel patterns and with the initial case occurring in Roseau. b) Simulated epidemic
using children's travel patterns and with the initial case occurring in Roseau.

This is not surprising, given that Roseau is the market center of the island and is
both the cultural and economic hub. Since travel is so frequent to Roseau (over 90
per cent of respondents visited at least once a month), even epidemics that begin
elsewhere reach the district almost immediately, which then provides a foothold
for the rest of the island.

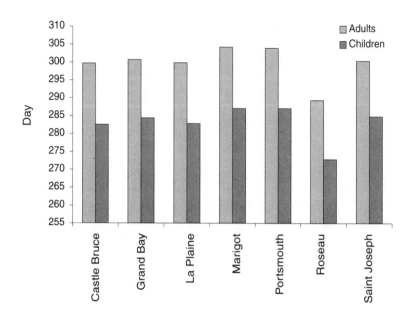

Figure 3.3 Comparison of peak times in simulations using adults' vs. children's mobility patterns. In all districts, epidemics peak earlier in simulations with children's mobility patterns.

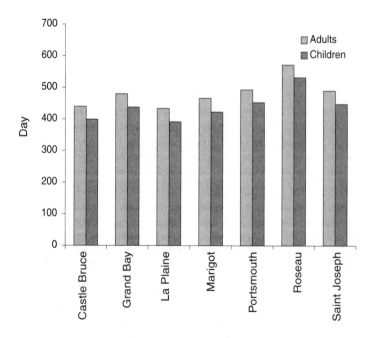

Figure 3.4 Comparison of end times in simulations using adults' vs. children's mobility patterns. In all districts, epidemics end earlier in simulations with children's mobility patterns.

Figure 3.3 shows that simulated epidemics using children's travel patterns peak earlier than comparable simulated epidemics using adults' travel patterns. Figure 3.4 shows that children's travel patterns also lead to epidemics that end earlier. Furthermore, simulated epidemics using both adults' and children's travel begin at about the same time (results not shown), which indicates that children's epidemics run their course more quickly than adult epidemics. This is related both to the fact that children travel less often, so epidemics within districts are reinforced from outside less frequently, and to the fact that children tend to spread their travel out among all districts while adults tend to localize their travel to particular districts, which has a tendency to slow spread throughout the island.

Simulation results are both similar to, and different from, patterns observed during the 1984 measles epidemic. Both simulations and the actual data result in some temporal separation of epidemic peaks across districts, although observed patterns are more marked than simulated epidemics. One major difference between the actual epidemic and simulated epidemics is the timing of the epidemic peak at Castle Bruce. This was the earliest peak in simulated epidemics, but the latest peak during the actual epidemic. Reasons for this difference are unclear, but may be related to reporting delays because of the remote location and relatively low socioeconomic and cultural status of the Castle Bruce district. It should also be noted that the 1984 measles epidemic is only one epidemic and is likely to be subject to chance factors. In addition, mobility patterns were estimated from data collected from 305 people sampled during one field season and may differ from patterns prevalent at other times. Furthermore, there is uncertainty in the values of some of the model parameters and differences between the simulation results and the observed epidemic may be related to inaccuracies in the values of these parameters. Comparisons of the simulations with the actual data, however, suggest that other actual epidemics are likely to be broadly similar to the 1984 measles epidemic, but may well differ in specific details.

The role of disease modeling in epidemiology and bioanthropology

Given that model simulations vary from observed data, it is natural to ask what the value is of a mathematical model. The answer to this question is that results from the analysis of a well-formulated mathematical model help to elucidate the relative importance of different possible factors affecting a system and stimulate the identification of new paths to take in future research. Although some of these results may appear common-sensical after the fact, often their importance is unrecognized at the outset. For example, an initial assumption of most people (including the author) would be that high rates of travel throughout a region would greatly increase the impact of an infectious disease epidemic within the region. Simulations of the spread of measles in Dominica and other diseases in other locations clearly indicate, however, that under a wide range of scenarios, mobility has a significant effect only on the introduction of a disease into a region, which influences the timing of epidemic peaks, but that once a disease enters a region other factors are far more important in determining the number of individuals in that region who are infected.

A second major use of mathematical modeling is that it provides a way to 'experiment' on human populations and allows a researcher to ask and generate answers to 'what-if' questions. This is especially important in studies of health and disease, because it is unethical to introduce a disease into a real population and observe how and why it spreads. In order for the results to be meaningful, however, it is essential to have well-designed models with parameter estimates that are grounded in reality. Experiments that vary parameter estimates in systematic ways can help researchers to understand whether an epidemiological event represents an unusual situation specific to a particular time and place or whether the event is an example of a more general pattern.

Infectious disease transmission involves biological and ecological interactions between a host, a pathogen, and often one or more intermediate species. The probability that transmission occurs is linked directly to biological variability in levels of resistance of a host, infectiousness of a pathogen, and many other factors. In addition, human disease transmission is profoundly influenced by individual behavior, social activities, and cultural norms and values. Therefore, the study of the transmission of infectious diseases in human populations is inherently bio-cultural. This combination of biology and culture that drives bioanthropological research also guarantees that studies of phenomena such as patterns of infectious disease epidemics are complex and difficult to understand fully. The techniques of mathematical modeling, which allow systematic exploration of the variety of biological and cultural factors that influence disease transmission, provide a powerful method to help simplify and understand the complexities of the real world.

Acknowledgements

I thank the Ministry of Health and people of the Commonwealth of Dominica for their support and encouragement, and Christopher Powell, Laura Supalla, Ryan Colvin, and Sara Blake for help in data collection and modeling activities. Funding for the Dominica project was provided by the National Science Foundation and the University of Missouri Research Council. Special thanks go to Nick Mascie-Taylor, Steve McGarvey, and Jean Peters who organized a fabulous symposium and edited volume, even if circumstances prevented those of us from overseas from attending.

References

Abdullah, A.S.M., Fielding, R. and Hedley, A.J. (2000) 'Hong Kong: an epicenter of increasing risk for HIV transmission? Overview and response', *AIDS Public Policy Journal*, 15: 4–16.

Ackerman, E., Elveback, L.R. and Fox, J.P. (1984) *Simulation of Infectious Disease Epidemics*, Springfield, IL: C.C. Thomas.

Anderson, R.M. and May, R.M. (1982) 'Coevolution of hosts and parasites', *Parasitology*, 85: 411–26.

Anderson, R.M. and May, R.M. (1996) 'The population biology of the interaction between HIV-1 and HIV-2: coexistence or competitive exclusion?', *AIDS*, 10: 1663–73.

Bartlett, M.S. (1961) 'Monte Carlo studies in ecology and epidemiology', *Proceedings of the 4th Berkeley Symposium on Mathematics, Statistics, and Probability*, 4: 39–55.

Barua, D. (1992) 'History of cholera', in D. Barua, and W.B. Greenough III (eds) *Cholera*, New York: Plenum, pp. 1–36.

Beaver, P.C. and Jung, R.C. (1985) *Animal Agents and Vectors of Human Disease*, 5th edn, Philadelphia: Lea and Febiger.

Birmingham, M.E., Lee, L.A., Ndayimirije, N., Nkurikiye, S., Hersh, B.S., Wells, J.G. and Deming, M.S. (1997) 'Epidemic cholera in Burundi: patterns of transmission in the Great Rift Valley Lake region', *Lancet*, 349: 981–5.

Blower, S.M. and Gerberding, J.L. (1998) 'Understanding, predicting, and controlling the emergence of drug-resistant tuberculosis: a theoretical framework', *J. Mol. Med.*, 76: 624–36.

Blower, S.M., Porco, T.C. and Darby, G. (1998) 'Predicting and preventing the emergence of antiviral drug resistance in HSV-2', *Nat. Med.* 4: 673–8.

Bonabeau, E., Toubiana, L. and Flahault, A. (1998) 'The geographical spread of influenza', *Proc. R. Soc. Lond. B Biol. Sci.*, 265: 2421–5.

Bradley, D.J. (1998) 'The influence of local changes in the rise of infectious disease', in B. Greenwood and K. De Cock (eds) *New and Resurgent Infections: Prediction, Detection and Management of Tomorrow's Epidemics*, Chichester: John Wiley and Sons, pp. 1–15.

Bull, J.J. (1994) 'Virulence', *Evolution*, 48: 1423–37.

Burnet, M. and White, D.O. (1972) *Natural History of Infectious Disease*, 4th edn, Cambridge: Cambridge University Press.

Burr, T., Hyman, J.M. and Myers, G. (2001) 'The origin of acquired immune deficiency syndrome: Darwinian or Lamarckian?', *Philos. Trans. R. Soc. Lond. B Biol. Sci.*, 356(1410): 877–87.

Cecere, M.C., Gurtler, R.E., Chuit, R. and Cohen, J.E. (1998) 'Factors limiting the domestic density of Triatoma infestans in northwest Argentina: a longitudinal study', *Bull. World. Health Organ.*, 76: 372–84.

Chin, J. (ed.) (2000) *Control of Communicable Diseases Manual*, 17th edn, Washington, DC: American Public Health Association.

Cogels, F.X., Coly, A. and Niang, A. (1997) 'Impact of dam construction on the hydrological regime and quality of a Sahelian Lake in the River Senegal basin', *Regul. River*, 13(1): 27–41.

Cohen, J.E. and Gurtler, R.E. (2001) 'Modeling household transmission of American trypanosomiasis', *Science*, 293: 694–8.

Coimbra, C.E.A., Jr. (1988) 'Human settlements, demographic pattern, and epidemiology in lowland Amazonia: the case of Chagas' disease', *Am. Anthropol.*, 90(1): 82–97.

Colwell, R.R. (1996) 'Global climate and infectious disease: the cholera paradigm', *Science*, 274: 2025–31.

Colwell, R.R. and Huq, A. (1994) 'Environmental reservoir of Vibrio cholerae: the causative agent of cholera', in M.E. Wilson, R. Levins and A. Spielman (eds) 'Disease in Evolution: Global Changes and Emergence of Infectious Diseases', *Ann. N. Y. Acad. Sci.*, 740: 44–54.

Dean, J.S., Gumerman, G.J., Epstein, J.M., Axtell, R.L., Swedlund, A.C., Parker, M.T. and McCarroll, S. (2000) 'Understanding Anasazi culture change through agent-based modeling', in T. Kohler and G. Gumerman (eds) *Dynamics in Human and Primate Societies: Agent-based Modeling of Social and Spatial Processes*, Oxford: Oxford University Press, pp. 179–205.

de Vlas, S.J., Van Oortmarssen, G.J., Gryseels, B., Polderman, A.M., Plaisier, A.P. and Habbema, J.D.F. (1996) 'SCHISTOSIM: a microsimulation model for the epidemiology and control of schistosomiasis', *Am. J. Trop. Med. Hyg.*, 55: 170–5.

Doran, J.E. and Palmer, M. (1995) 'The EOS project: integrating two models of Paleolithic social change' in N. Gilbert and R. Conte (eds) *Artificial Societies: The Computer Simulation of Social Life*, London: UCL Press, pp. 103–25.

Doran, J.E., Palmer, M., Gilbert, N. and Mellars, P. (1994) 'The EOS project: modeling Upper Paleolithic social change', in N. Gilbert and J. Doran (eds) *Simulating Societies: The Computer Simulation of Social Phenomena*, London: UCL Press, pp. 195–221.

Dubos, R. (1965) *Man Adapting*, New Haven, CT: Yale University Press.

Epstein, J.M. and Axtell, R. (1996) *Growing Artificial Societies: Social Science from the Bottom Up*, Washington, DC: Brookings Institution Press.

Epstein, P.R. (1999) 'Climate and health', *Science*, 25: 347–8.

Ewald, P.W. (1994) *Evolution of Infectious Disease*, Oxford: Oxford University Press.

Ewald, P.W. (1996) 'Guarding against the most dangerous emerging pathogens: insights from evolutionary biology', *Emer. Infect. Dis.*, 2: 245–57.

Ewald, P.W. (1998) 'The evolution of virulence and emerging diseases', *J. Urban Health*, 75: 480–91.

Fix, A.G. (1984) 'Kin groups and trait groups: population structure and epidemic disease selection', *Am. J. Phys. Anthropol.*, 65: 201–12.

Friedman, J.F., Kurtis, J.D., McGarvey, S.T., Fraga, A.L., Silveira, A., Pizziolo, V., Gazzinelli, G., LoVerde, P. and Correa-Oliveira, R. (2001) 'Comparison of self-reported and observed water contact in an S. mansoni endemic village in Brazil', *Acta Trop.*, 78: 251–9.

Garrett, L. (1994) *The Coming Plague: Newly Emerging Diseases in a World Out of Balance*, New York: Penguin.

Gilbert, N. and Conte, R. (1995) *Artificial Societies: The Computer Simulation of Social Life*, London: UCL Press.

Gurtler, R.E. (1999) 'Control campaigns against Triatoma infestans in a rural community of Northwestern Argentina', *Medicina (B. Aires)*, 59: 47–54.

Habbema, J.D.F., de Vlas, S.J., Plaisier, A.P. and Van Oortmarssen, G.J. (1996) 'The microsimulation approach to epidemiologic modeling of helminthic infections, with special reference to schistosomiasis', *Am. J. Trop. Med. Hyg.*, 55: 165–9.

Hammad, T.A., Abdel Wahab, M.F., De Claris, N., El Sahly, A., El Kady, N. and Strickland, G.T. (1996) 'Comparative evaluation of the use of artificial neural networks for modelling the epidemiology of Schistosomiasis mansoni', *Trans. R. Soc. Trop. Med. Hyg.*, 90: 372–6.

Huston, C.D. and Petri, W.A. (2001) 'Emerging and reemerging intestinal protozoa', *Curr. Opin. Gastroenterol.*, 17: 17–23.

Kiszewski, A.E. and Spielman, A. (1994) 'Virulence of vector-borne pathogens – a stochastic automata model of perpetuation', in M.E. Wilson, R. Levins and A. Spielman (eds) 'Disease in Evolution: Global Changes and Emergence of Infectious Diseases', *Ann. N. Y. Acad. Sci.*, 740: 249–59.

Kohler, T.A. (2000) 'Putting social sciences together again: an introduction to the volume', in T. Kohler and G. Gumerman (eds) *Dynamics in Human and Primate Societies: Agent-based Modeling of Social and Spatial Processes*, Oxford: Oxford University Press, pp. 1–18.

Kohler, T.A., Kresl, J., Van West, C., Carr, E. and Wilshusen, R.H. (2000) 'Be there then: a modeling approach to settlement determinants and spatial efficiency among Late Ancestral Pueblo populations of the Mesa Verde region, U.S. Southwest', in T. Kohler and G. Gumerman (eds) *Dynamics in Human and Primate Societies: Agent-based Modeling of Social and Spatial Processes,* Oxford: Oxford University Press, pp. 145–78.

Kretzschmar, M. and Morris, M. (1996) 'Measures of concurrency in networks and the spread of infectious disease', *Math. Biosci.,* 133: 165–95.

Lake, M.W. (2000) 'MAGICAL computer simulation of Mesolithic foraging', in T. Kohler and G. Gumerman (eds) *Dynamics in Human and Primate Societies: Agent-based Modeling of Social and Spatial Processes,* Oxford: Oxford University Press, pp. 107–43.

Levin, B.R. (1996) 'The evolution and maintenance of virulence in microparasites', *Emerg. Infect. Dis.,* 2: 93–102.

Levin, B.R., Lipsitch, M., Perrot, V., Schrag, S., Antia, R., Simonsen, L., Walker, N.M. and Stewart, F.M. (1997) 'The population genetics of antibiotic resistance', *Clin. Infect. Dis.,* 24: S9–S16.

Levins, R., Awerbuch, T., Brinkmann, U., Eckardt, I., Epstein, P., Makhoul, N., Albuquerque de Possas, C., Puccia, C., Spielman, A. and Wilson, M.E. (1994) 'The emergence of new diseases', *Am. Sci.,* 82: 52–60.

Lin, J., Andreasen, V. and Levin, S. (1999) 'Dynamics of influenza A drift: the linear three-strain model', *Math. Biosci.,* 12: 33–51.

Lindblade, K.A., Walker, E.D., Onapa, A.W., Katungu, J. and Wilson, M.L. (1999) 'Highland malaria in Uganda: prospective analysis of an epidemic associated with El Nino', *Trans. R. Soc. Trop. Med. Hyg.,* 93: 480–7.

Lindblade, K.A., Walker, E.D., Onapa, A.W., Katungu, J. and Wilson, M.L. (2000) 'Land use change alters malaria transmission parameters by modifying temperature in a highland area of Uganda', *Trop. Med. Int. Health.,* 5: 263–74.

Lindsay, S.W. and Birley, M.H. (1996) 'Climate change and malaria transmission', *Ann. Trop. Med. Parasitol.,* 90: 573–88.

Lindsay, S.W. and Martens, W.J.M. (1998) 'Malaria in the African highlands: past, present, and future', *Bull. World Health Organ.,* 76: 33–45.

Lipsitch, M. and Levin, B.R. (1998) 'Population dynamics of tuberculosis treatment: mathematical models of the roles of non-compliance and bacterial heterogeneity in the evolution of drug resistance', *Int. J. Tuberc. Lung Dis.,* 2: 187–99.

Lipsitch, M. and Nowak, M.A. (1995) 'The evolution of virulence in sexually transmitted HIV/AIDS', *J. Theor. Biol.,* 174: 427–40.

Longini, I.M., Jr. (1988) 'A mathematical model for predicting the geographic spread of new infectious agents', *Math. Biosci.,* 90: 367–83.

Martens, W.J.M., Jetten, T.H. and Focks, D.A. (1997) 'Sensitivity of malaria, schistosomiasis, and dengue to global warming', *Climatic Change,* 35: 145–56.

Massad, E. and Forattini, O.P. (1998) 'Modelling the temperature sensitivity of some physiological parameters of epidemiologic significance', *Ecosyst. Health,* 4: 119–29.

Mata, L. (1994) 'Cholera El Tor in Latin America, 1991–1993', in M.E. Wilson, R. Levins, and A. Spielman (eds) *Disease in Evolution, Global Changes and Emergence of Infectious Diseases,* The New York Academy of Sciences, New York, pp. 55–68.

May, R.M. (1983) 'Parasitic infections as regulators of animal populations', *Am. Sci.,* 71: 36–45.

May, R.M. and Anderson, R.M. (1983) 'Parasite–host coevolution' in D.J. Futuyma, and M. Slatkin (eds), *Coevolution,* Sunderland, MA: Sinauer, pp. 186–206.

McGrath, J.W. (1988) 'Social networks of disease spread in the Lower Illinois Valley: a simulation approach', *Am. J. Phys. Anthropol.*, 77: 483–96.

Milgram, S. (1967) 'The small world problem', *Psychol. Today*, 2: 60–7.

Moore, C. and Newman, M.E.J. (2000) 'Epidemics and percolation in small-world networks', *Phys. Rev. E*, 61: 5678–82.

Morris, M. and Kretzschmar, M. (1997) 'Concurrent partnerships and the spread of HIV', *AIDS*, 11: 641–8.

Mouchet, J., Manguin, S., Sircoulon, J., Laventure, S., Faye, O., Onapa, A.W., Carnevale, P., Julvez, J. and Fontenille, D. (1998) 'Evolution of malaria in Africa for the past 40 years: impact of climatic and human factors', *J. Am. Mosq. Control Assoc.*, 14: 121–30.

Murray, G.D. and Cliff, A.D. (1975) 'A stochastic model for measles epidemics in a multi-region setting', *T. I. Br. Geogr.*, 2: 158–74.

Nakazawa, M., Ohmae, H., Ishii, A., Leafasia, J. (1998) 'Malaria infection and human behavioral factors: a stochastic model analysis for direct observation data in the Solomon Islands', *Am. J. Hum. Biol.*, 10: 781–9.

Ostroff, S.M. and Kozarsky, P. (1998) 'Emerging infectious diseases and travel medicine', *Infect. Dis. Clin. North Am.*, 12: 231–41.

Palkovich, A.M. (1981) 'Demography and disease patterns in a protohistoric Plains group: a study of the Mobridge site (39WW1)', *Plains Anthropol.*, 26 (Memoir 17): 71–84.

Panetta, J.C. (1997) 'A logistic model of periodic chemotherapy with drug resistance', *Appl. Math. Lett.*, 10: 123–7.

Pease, C.M. (1987) 'An evolutionary epidemiological mechanism, with applications to type A influenza', *Theor. Pop. Biol.*, 31: 422–52.

Pool, I. and Kochen, M. (1978) 'Contacts and influence', *Soc. Networks*, 1: 1–48.

Ramenofsky, A.F. (1987) *Vectors of Death*, Albuquerque: University of New Mexico Press.

Reiter, P. (1998) 'Aedes albopictus and the world trade in used tires, 1988–1995: the shape of things to come?', *J. Am. Mosq. Control Assoc.*, 14: 83–94.

Reiter, P. (2001) 'Climate change and mosquito-borne disease', *Environ. Health Perspect.*, 109: 141–61.

Russell, S.J. and Norvig, P. (1995) *Artificial Intelligence: A Modern Approach*, Englewood Cliffs, NJ: Prentice-Hall.

Rvachev, L.A. and Longini, I.M., Jr. (1985) 'A mathematical model for the global spread of influenza', *Math. Biosci.*, 75: 3–22.

Sattenspiel, L. and Dietz, K. (1995) 'A structured epidemic model incorporating geographic mobility among regions', *Math. Biosci.*, 128: 71–91.

Smallman-Raynor, M. and Cliff, A.D. (1998a) 'The Philippines insurrection and the 1902–4 cholera epidemic: Part I – Epidemiological diffusion processes in war', *J. Hist. Geogr.*, 24: 69–89.

Smallman-Raynor, M. and Cliff, A.D. (1998b) 'The Philippines insurrection and the 1902–4 cholera epidemic: Part II – Diffusion patterns in war and peace', *J. Hist. Geogr.*, 24: 188–210.

Smallman-Raynor, M. and Cliff, A.D. (1999) 'The spatial dynamics of epidemic diseases in war and peace: Cuba and the insurrection against Spain, 1895–98', *T. I. Brit. Geogr.*, 24: 331–52.

Smallman-Raynor, M. and Cliff, S.D. (2001) 'Epidemic diffusion processes in a system of US military camps: transfer diffusion and the spread of typhoid fever in the Spanish–American War, 1898', *Ann. Assoc. Am. Geogr.*, 91: 71–91.

Stewart, F.M., Antia, R., Levin, B.R., Lipsitch, M. and Mittler, J.E. (1998) 'The population genetics of antibiotic resistance II: analytic theory for sustained populations of bacteria in a community of hosts', *Theor. Pop. Biol.*, 53: 152–65.

Stilianakis, N.I., Perelson, A.S. and Hayden, F.G. (1998) 'Emergence of drug resistance during an influenza epidemic: insights from a mathematical model', *J. Infect. Dis.*, 177: 863–73.

Tanguy, J.C., Ribiere, C., Scarth, A. and Tjetjep, W.S. (1998) 'Victims from volcanic eruptions: a revised database', *Bull. Volcanol.*, 60: 137–44.

Thomas, R. (2000) 'Reconstructing the space–time structure of the HIV/AIDS epidemic for the countries of Western Europe', *T. I. Brit. Geogr.*, 25: 445–63.

Thomas, R. and Smith, T.E. (2000) 'Multiregion contact systems for modelling STD epidemics', *Stat. Med.*, 19: 2479–91.

Thornton, R., Miller, T. and Warren, J. (1991) 'American Indian population recovery following smallpox epidemics', *Am. Anthropol.*, 93: 28–45.

Upham, S. (1986) 'Smallpox and climate in the American Southwest', *Am. Anthropol.*, 88: 115–28.

Watts, D.J. (1999) 'Networks, dynamics, and the small-world phenomenon', *American Journal of Sociology*, 105: 493–527.

Watts, S.J. (1987) 'Population mobility and disease transmission: the example of guinea worm', *Soc. Sci. Med.*, 25: 1073–81.

Wawer, M.J., Podhisita, C., Kanungsukkasem, U., Pramualratana, A. and McNamara, R. (1996) 'Origins and working conditions of female sex workers in urban Thailand: consequences of social context for HIV transmission', *Soc. Sci. Med.*, 42: 453–62.

Whitmore, T.M. (1992) *Disease and Death in Early Colonial Mexico: Simulating Amerindian Depopulation*, Boulder, CO: Westview.

Wills, C. (1996) *Yellow Fever, Black Goddess: The Coevolution of People and Plagues*, Reading, MA: Addison-Wesley.

Wilson, M.E., Levins, R. and Spielman, A. (eds) (1994) *Disease in Evolution: Global Changes and Emergence of Infectious Diseases*, The New York Academy of Sciences, New York.

World Health Organization (1996) *The World Health Report 1996: Fighting Disease, Fighting Development*, Report of the Director-General, Geneva, Switzerland.

Yang, H.M. and Ferreira, M.U. (2000) 'Assessing the effects of global warming and local social and economic conditions on malaria transmission', *Rev. Saude. Publica*, 34: 214–22.

Zeller, H.G. (1998) 'Dengue, arboviruses and migrations in the Indian Ocean', *Bull. Soc. Pathol. Exot.*, 91(1): 56–60.

Part II
Current challenges

4 Genetic epidemiology of parasitic diseases

Sarah Williams-Blangero, John L. VandeBerg and John Blangero

Introduction

Parasitic diseases persist as major global health threats despite the dramatic advances in medical science that have occurred in the last 100 years. For some parasitic diseases, the prevalence today is similar to that observed 50 years ago even though new effective pharmacological treatments have been developed (Chan *et al.*, 1994; Chan, 1997). The global health burden associated with these diseases is enormous. For example, using the disability adjusted life years (DALYs) measure of Murray (1994), it has been estimated that 39.0 million years of healthy life were lost worldwide due to soil-transmitted intestinal worm infections (hookworm, roundworm, and whipworm) and 35.7 million years to malaria in 1990 (Chan, 1997). By comparison, diabetes only accounted for 8.0 million years of healthy life lost in that same year, while motor vehicle accidents accounted for 31.7 million DALYs (Chan, 1997). Clearly there is a significant need for continued research aimed at identifying new mechanisms to more effectively prevent, treat, and cure parasitic diseases.

Genetic epidemiology and parasitic diseases

Genetic epidemiology is one of the scientific fields identified by Anthony Fauci, Director of the National Institute of Allergy and Infectious Diseases of the U.S. National Institutes of Health, as being critical for progress in infectious disease research in the 21st century, particularly in the area of host susceptibility (Fauci, 2001). Genetic epidemiological approaches are used to characterize the host genetic components influencing disease susceptibility and outcome. While these techniques have been commonly applied in studies of complex noninfectious diseases such as atherosclerosis, diabetes, and obesity, they have been applied to infectious diseases, and particularly parasitic diseases, relatively infrequently (but see Abel *et al.*, 1991, 1992; Marquet *et al.*, 1996; Williams-Blangero *et al.*, 1997a, 1997b, 1999, 2002a, 2002b; Garcia *et al.*, 1999).

The identification and characterization of individual genes influencing disease are the ultimate goals of genetic epidemiological studies of parasitic infections. These studies can facilitate drug discovery through identification of new mechanisms influencing susceptibility to, or progression of, disease that can be targeted in the drug development process (Dykes, 1996; Gelbert and Gregg 1997).

Genetic epidemiological studies may be particularly informative for parasitic diseases for which there is no cure, no acquired immunity, and/or for which there is variation in response to infection. For example, Chagas' disease (American trypanosomiasis), which is caused by infection with the protozoan *Trypanosoma cruzi*, has no effective pharmacological prevention or cure. Identifying the genes influencing susceptibility to infection with the parasitic organism or progression of the resulting disease may suggest new targets for drug development efforts aimed at treatment and prophylaxis.

Genetic approaches may be useful for characterizing highly susceptible individuals when there is no acquired immunity to the parasite and reinfection can easily occur. For example, there is no evidence for development of acquired immunity to *Ascaris lumbricoides*, an intestinal worm infection that afflicts 1,274 million people (Silva *et al.*, 1997a) throughout the world. Individuals may be readily reinfected after treatment (Seltzer and Barry, 1999). In this case, genetic information may be useful for both targeting drugs to the most susceptible portion of the population and suggesting new biological mechanisms to target in the development of a vaccine or more effective anthelmintics.

There are many parasitic infections that show varied outcomes in different individuals and there may be a genetic basis to this variation in disease progression. Infection by a parasitic organism may have relatively benign effects in some individuals, while having a severe outcome in others. Variability in disease outcome is characteristic of infections by a broad range of parasitic organisms. For example, *T. cruzi* infection can be asymptomatic or lead to a long-term chronic disease which may or may not be fatal. Similarly, there is considerable variation in the level of anaemia associated with hookworm (*Necator americanus* and *Ancylostoma duodenale*). Genetic epidemiological approaches can be used to determine if there are genes influencing differential response to infection. If genetic factors are involved, identifying the specific genes responsible can provide a mechanism for identifying the individuals most likely to have a poor outcome, and again allow for targeting treatment to the individuals who need it most.

The parasitic disease in genetic epidemiological studies of parasitic diseases

Many of the parasitic diseases that represent a large proportion of the global health burden are common in endemic populations, with 10 per cent or more of the population experiencing infection. For genetic analysis of a parasitic disease with a frequency greater than 10 per cent, the infection ideally should have several characteristics. First, it should be possible (financially and practically) to assess disease status in the large number of individuals required to have an adequate sample size for genetic analysis. Second, there should be a quantitative measure of disease intensity. While it is possible to analyze genetically a qualitative assessment of disease status (e.g. infected versus uninfected) by assuming an underlying distribution of liability, the analysis of quantitative disease measures inherently has greater statistical power. Finally, there should be universal or measurable risk of exposure to disease. Many of the parasitic diseases have these characteristics, making them ideal candidates for genetic epidemiological study and possible models for understanding the genetics of other complex infectious diseases.

The host population in genetic epidemiological studies of parasitic diseases

The statistical power for genetic epidemiological analysis of common diseases is a function of the size and complexity of the pedigree (Blangero *et al.*, 2000; Dyer *et al.*, 2001). When the disease of interest has a frequency of greater than 10 per cent in the population, extended pedigrees that are not selected with respect to the parasitic disease phenotype will be informative for genetic analyses of both the disease trait and any other disease-related or normal phenotypes that are assessed in the population.

In geographically limited, stable populations with low migration rates, extended pedigrees can be reconstructed quite easily from data gathered through house-to-house sampling. Simply obtaining information on mother's identity, father's identity, mother's mother's identity, mother's father's identity, father's mother's identity, and father's father's identity for each individual resident in each household, and assigning unique identifier numbers to each individual, allows for pedigree reconstruction from population-based data. PEDSYS, a pedigree-based data management system, can be used to reconstruct pedigrees from such data (Dyke, 1989). This approach has been used to reconstruct a single pedigree containing all 1261 individuals sampled for a study of intestinal worm infections (Williams-Blangero *et al.*, 1999).

In less stable populations, pedigrees can be constructed around individuals (probands) selected without regard to disease phenotype, but on the basis of age and number of pedigree members available for sampling. For example, if the study is restricted to adults, participants may be restricted to individuals at least 18 years of age who have at least two siblings residing locally and living parents who are also available for sampling. Such a pedigree construction approach has been used with success to develop extended pedigrees in urban areas of the United States (e.g. Mitchell *et al.*, 1996; MacCluer *et al.*, 1999).

Large extended pedigrees provide statistical power for detection of genetic effects. However, parasitic diseases also frequently have strong environmental components to disease susceptibility. Thus, the optimal sampling scheme for a genetic study of parasitic diseases requires that you are able to distinguish both genetic effects and local environmental effects. The ideal sampling scheme involves sampling large pedigrees over multiple households. The large pedigrees provide the power to detect genetic effects, while the presence of multiple households within pedigrees allows discrimination of local environmental effects.

Genetic epidemiological analysis

Quantitative genetic approaches assess how much of the variation in the trait is attributable to genetic factors; this proportion is termed the heritability (h^2) of the trait. A quantitative genetic analysis is a useful first step in assessing the role of genetics in determining susceptibility to parasitic infections. Quantitative genetic approaches have been applied to a number of different parasitic diseases. For example, the heritability of hookworm infection has been assessed in a population from rural Zimbabwe, with approximately 37 per cent of the variation in worm load as measured by eggs per gram of faeces attributable to genetic factors (Williams-Blangero et al., 1997a). Similarly, 28 per cent of the variation in susceptibility to whipworm (*Trichuris trichiura*) is attributable to genetic factors in both a Chinese population and a Nepalese population that differ markedly in the prevalence of infection (86 per cent in the Chinese population and 14 per cent in the Nepalese population) (Williams-Blangero et al., 2002a). However, a persistent problem with quantitative genetic approaches is that it is impossible to completely disentangle environmental and genetic effects.

Genome scans allow unambiguous chromosomal localization and identification of genetic effects because it is impossible for the effects of environmental factors to segregate throughout an extended pedigree in the same way as genetic factors. A genome scan typically involves typing each of 500–1000 individuals for approximately 400 genetic markers spaced evenly at 10 centimorgan intervals throughout the genome. Linkage methods are then used to determine if the disease trait co-segregates through the pedigree with the genetic marker (Blangero et al., 2000). If the disease trait is linked to a specific genetic marker, then the chromosomal location of the gene influencing the disease trait can be determined. Localization is the first step towards gene identification. There are few genome scans for parasitic diseases ongoing, and only one has reported linkage for a parasitic disease. Marquet and colleagues (1996) published this first linkage result for schistosomiasis, localizing a gene influencing intensity of infection with *Schistosoma mansoni* to chromosome 5.

T. cruzi infection as an example of a parasitic disease with no chemoprevention or cure

Chagas' disease is found throughout South and Central America (PAHO, 1982; Schmunis, 1991). It results from infection with the protozoan *T. cruzi* which is transmitted to humans by triatomid bugs. In endemic populations, infection occurs during childhood. Triatomines feed on a broad range of mammals, including opossums, sheep, guinea pigs, dogs, and cats, which are reservoirs for this zoonotic disease (Garcia Zapata et al., 1986; Gurtler et al., 1992). The goal of completely eliminating the vector is unlikely to be achieved given its broad geographic range and the diversity of animals on which it feeds. An added problem associated with eliminating the vector is the development of insecticide resistance in triatomines (Zerba, 1999; Vassena et al., 2000).

Chagas' disease has both an acute phase and a chronic phase. The acute phase immediately following infection lasts two to three months and is sometimes characterized by an inflammation known as a Chagoma at the site of infection (Nogueira and Coura, 1990). The infection may be completely benign during the acute phase, although 5 to 10 per cent of infected individuals experience severe and potentially fatal disease in this phase (Nogueira and Coura, 1990). About 40 per cent of infected individuals remain seropositive but disease-free following the acute phase. In the remaining 60 per cent of cases, the disease may be quiescent for as long as 30 years until the chronic phase characterized by progressive cardiomyopathy, megacolon, or megaesophagus begins (Morris *et al.*, 1990; Nogueira and Coura, 1990).

While transmission has been reduced in some areas due to active vector control programs, the disease persists as the leading cause of heart disease in South and Central America and remains an international public health concern (Tanowitz *et al.*, 1992; Schofield and Dias, 1999). Chagas' disease is of increasing significance in North America (e.g. Ochs *et al.*, 1996; Di Pentima *et al.*, 1999). *T. cruzi* has been isolated from naturally ranging triatomine bugs and mammals captured across the southern and southwestern U.S. (Navin *et al.*, 1985; Beard *et al.*, 1988; Yaeger, 1988; Bradley *et al.*, 2000; Yabsley and Noblet, 2002). The increase in Latin American immigrants to the U.S. has raised concerns about the implications of *T. cruzi* for blood and tissue banks (Kirchhoff and Neva, 1985; Kirchhoff, 1989, 1993a, 1993b; Schmunis, 1991). Transfusion associated- (Nickerson *et al.*, 1989) and natural-transmission (Navin *et al.*, 1985) of *T. cruzi* have occurred in North America. More recently, three cases of transmission of *T. cruzi* infection in the U.S. resulted from organ transplants from a single infected donor (Zayas *et al.*, 2002).

There are no vaccines and no effective prophylactic drugs for *T. cruzi* infection; treatment of the disease is still highly problematic (Kinnamon *et al.*, 1997; Fairlamb, 1999; Stoppani, 1999). The two drugs used against this infection are both carcinogenic in rabbits and are effective in reducing parasitemia only if given shortly after infection (Teixeira *et al.*, 1990a, 1990b; Kirchhoff, 1999). The identification of the specific genetic factors influencing this complex disease entity could yield important new information about biological mechanisms to be targeted for prevention and intervention in this disease process that now affects over 16 million individuals (WHO, 1991, 1993, 1999) and threatens an additional 120 million individuals (WHO, 1991, 1993, 1999). The lack of an adequate pharmacological intervention makes Chagas' disease an ideal candidate for a novel drug discovery strategy based on genome analysis of the human host's susceptibility to infection with *T. cruzi* and to progression of the disease subsequent to infection.

It has been speculated that genetic factors might be involved in determining differential susceptibility to *T. cruzi* infection and to disease progression in Chagas' disease (Teixeira, 1979; Nogueira and Coura, 1990). Feitosa and Krieger (1993) created a mathematical model based on a double binomial with one tail excess to account for the clustering of Chagas' disease in northeastern Brazil. The model suggested that disease was concentrated in a small proportion of families that were at high risk for infection. This observation is consistent with a genetic determination of susceptibility, although the authors attributed the clustering to environmental factors. A case control study of sibling history of heart disease in Chagas' disease patients and normal individuals found evidence for familial aggregation of the disease (Zicker *et al.*, 1990). Associations between HLA haplotypes and Chagas' disease have been reported in a number of studies (Llop *et al.*, 1991; Fernandez-Mestre *et al.*, 1998; Colorado *et al.*, 2000; Nieto *et al.*, 2000).

To assess the genetics of susceptibility to *T. cruzi* infection and its disease consequences, the Posse Family Health Study was initiated in 1996. This large-scale genetic epidemiological study of Chagas' disease is based in the Posse region of the state of Goiás in Brazil. Over 60 per cent of the adults in the region are seropositive for infection with *T. cruzi* (Williams-Blangero *et al.*, 1997b). The first step was to assess whether or not there was evidence for the influence of genetic factors on susceptibility to infection. Seronegative long-term residents are found in all endemic areas, despite their prior exposure to *T. cruzi* infected triatomine bugs. This suggests that there may be differential susceptibility to becoming infected with *T. cruzi*.

Unlike many infectious agents which are cleared from the body by the immune system, *T. cruzi* persists throughout the life of the host. There are no documented cases of spontaneous clearance of *T. cruzi* from infected individuals, nor of antibody titers diminishing to undetectable levels. Therefore, for *T. cruzi* infection, sero-positivity status (positive or negative) reflects whether or not an individual ever became infected with the parasite (and still is infected).

To test the hypothesis that genetic factors mediate variation in seropositivity status, quantitative genetic approaches were used to determine how much of the variation in the trait was attributable to genetic factors. Seropositivity data gathered from 716 adults residing in the Posse region was analyzed (Williams-Blangero *et al.*, 1997b). Individuals were sampled through house-to-house surveys, and genealogical information, including age, sex, parental names, grandparental names, and residences, were collected for all persons living in the sampled households.

These demographic and genealogical data allowed construction of extended multihousehold pedigrees using the pedigree-based data management system PEDSYS (Dyke, 1989). Individuals were defined as belonging to a pedigree if they were biologically related to anyone else in the pedigree or if they were unrelated marry-ins who had not yet had children in the pedigree, but lived in the same house as their spouse who was biologically related to someone else in the pedigree. The unrelated marry-ins were included primarily to improve estimation of common household effects. Five hundred and twenty-five of the 716 individuals belonged to a total of 146 pedigrees composed of between two and 103 individuals. In addition to the 525 people who could be assigned to pedigrees, there were 191 unrelated individuals who were retained in the analysis to improve parameter estimation. The sample included 418 households, 371 of which had one or two sampled occupants, 44 of which had three or four sampled occupants, and three for which five or six occupants were sampled. Seropositivity for *T. cruzi* infection was determined from a finger prick sample (Williams-Blangero *et al.*, 1997b).

To estimate the proportion of the seropositivity trait variation attributable to genes, common environment, and random individual-specific environmental effects, an adaptation of pedigree-based maximum likelihood variance decomposition methods was used (Hopper and Mathews 1982; Lange and Boehnke 1983) as implemented in FISHER (Lange *et al.*, 1988) that utilized all pedigree and household sharing information. Because the seropositivity trait is dichotomous, inferences were made based on an underlying liability scale where seropositivity was viewed as a threshold point on a distribution of liability to become seropositive, i.e. the distribution of susceptibility to being infected (Williams-Blangero *et al.*, 1997b).

The quantitative genetic analysis of seropositivity in the Posse sample indicated that 56 per cent of the variation in susceptibility to becoming seropositive was attributable to genetic factors ($h^2 = 0.56 \pm 0.27$) (Williams-Blangero *et al.*, 1997b). A further 23 per cent of the variation was attributable to common household effects, indicating that there is a significant environmental component to this vector-borne disease (Williams-Blangero *et al.*, 1997b). However, the household reflects the current environment rather than the household in which the person lived at the time of infection. Therefore, the current household may be a proxy measure of socioeconomic status or some similar environmental factor correlated with the childhood household environment in which the person was first exposed to risk of infection. The evidence for substantial genetic influences on susceptibility to *T. cruzi* infection in this population led to the development of a genome scan that will ultimately involve 1500 individuals in large extended pedigrees who will each be typed for 375 markers, and characterized for seropositivity and a range of phenotypes associated with progression of Chagas' disease.

The genome scan approach allows clear discrimination between genetic effects and environmental effects. A genome scan was initiated in the Posse Family Health Study, and a small subset of the sample was used for preliminary linkage analyses. Ninety-nine markers in 162 individuals from two large families were typed. Linkage analysis was then performed to localize genes influencing the seropositivity trait. The sample included 72 males and 90 females aged between 18 and 79 years. The seroprevalence rate was 63 per cent in the two pedigrees. Marker data was generated on 51 markers distributed evenly at 10 centimorgan intervals across three chromosomes (12 markers on chromosome 1, 20 markers on chromosome 5, and 19 markers on chromosome 6). These chromosomes were selected for the initial analyses because of the presence of multiple candidate genes. Variance component linkage analysis as implemented in the program SOLAR (Almasy and Blangero, 1998) was used to assess the evidence for linkage between the seropositivity trait and each of the 51 available genetic markers.

Preliminary analyses of this small data set revealed suggestive linkages on two chromosomes. The strongest signal yielded a LOD score of 2.28 ($p = 0.0006$) near STR marker D5S408 on chromosome 5. This region (located towards the q terminus) has an important candidate gene that is a member of the T-cell receptor alpha-chain family and is involved in the regulation of T-cells. The second strongest signal was found on chromosome 1 between markers D1S1679 and D1S1677 in region 1q21–q22. This area produced a LOD score of 1.54 ($p = 0.0039$). The peak LOD score was within 2.5 centimorgans of another immune function related candidate gene *CD48* which encodes for the CD48 antigen. Although little is known about its function, it is probably a cell adhesion molecule involved in regulating B- and T-cell activation (Klyushnenkova *et al.*, 1996; Ianelli *et al.*, 1997; Tissot *et al.*, 1997). Chromosome 6 was examined because of the suggestive evidence of previous HLA association studies, but no evidence for linkage to the HLA gene family was found.

The preliminary genetic results from the Posse Family Health study are exciting because they indicate that seropositivity for *T. cruzi* infection has a substantial heritable component whose genetic determinants can be mapped using genome scanning. Although preliminary, the quantitative genetic and linkage results provide support for the influence of individual genetic factors on differential susceptibility to *T. cruzi* infection. The ultimate identification of these genes may suggest new mechanisms to target in pharmacological approaches to this incurable parasitic infection.

Ascariasis as an example of a treatable parasitic disease with no acquired immunity

Infection with roundworm (*Ascaris lumbricoides*) remains a major international health concern, affecting more than a quarter of the world's population (Tanowitz *et al.*, 1994; Silva *et al.*, 1997a; Holland and Kennedy, 2002). While individuals with low roundworm loads may be asymptomatic, heavily infected cases can experience serious morbidity and even mortality due to intestinal blockage (Chrungroo *et al.*, 1992; Silva *et al.*, 1997a, 1997b). Chronic ascariasis has been implicated in the development of malnutrition in children and consequent deficits in growth and development (Stephenson, 1987; Gupta, 1990; Crompton, 1992; Silva *et al.*, 1997a; Stephenson *et al.*, 2000; Towne *et al.*, 2000). It has been estimated that 1274 million people are infected worldwide, that 59 million are at risk of serious health deficits associated with the disease, that 200,000 individuals will develop serious life-threatening consequences each year, and that approximately 10,000 individuals per year will die of the consequences of infection (Silva *et al.*, 1997a).

Ascaris lumbricoides is a parasitic nematode, the adult form of which lives in the human intestine for between 6 and 18 months (Seltzer and Barry, 1999). Female worms grow to approximately 40 cm in length, while males achieve a maximum length of about 35 cm (Seltzer and Barry, 1999). Eggs are passed through the faeces and become infective after embryonating in the soil (Seltzer and Barry, 1999). Infection occurs through ingestion of contaminated soil such as when unwashed fruits and vegetables are consumed. Roundworms do not multiply within the host, so each worm represents a separate infection; thus, wormload is a quantitative measure of infection.

Even with the availability of effective anthelmintics (Silva *et al.*, 1997c), infection levels can remain high because of the ubiquity of exposure and the resilience of *Ascaris* eggs (Seltzer and Barry, 1999). There is no acquired immunity to *A. lumbricoides* and individuals are readily reinfected after treatment. The familial nature of roundworm infection has been noted (Williams *et al.*, 1974; Chai *et al.*, 1983; Forrester *et al.*, 1988), and several studies have implicated genetic factors in determining this familial aggregation (Holland *et al.*, 1992; Ramsay *et al.*, 1999).

The Jiri Helminth Project was established in 1995 in the Jiri region of eastern Nepal to assess the genetic components influencing susceptibility to helminthic infection (Williams-Blangero and Blangero, 2002). The study population is the Jirel endogamous ethnic group which has been the subject of genetic research since 1985 (Williams-Blangero, 1990). As a result of the 17 years of research conducted with the members of the Jirel population, a wealth of pedigree information exists, allowing construction of one of the most extensive and complete human pedigrees available for genetic analysis. The current 1261 participants in the Jiri Helminth Project belong to a single pedigree that contains over 26,000 relationship pairs (from parent–offspring pairs to ninth degree relative pairs) that are informative for genetic analysis (Williams-Blangero *et al.*, 1999, 2002b). The Jirel pedigree includes 250 households which provide substantial statistical power for detecting common environmental effects. This is a tremendously powerful pedigree structure for genetic epidemiological analyses of the helminthic infections prevalent in the area.

Ascaris worm loads were assessed by a number of measures. Egg counts were determined on small faecal samples collected prior to anthelminthic treatment, total worm counts were determined from complete stools collected for 96 hours following treatment with the anthelminthic albendazole, and total worm weight was determined as the sum of the weights of all roundworms collected from a given individual in the 96-hour collection period.

Approximately 30 per cent of the Jirel population is infected with *Ascaris* (Williams-Blangero *et al.*, 1999). As is typical for many helminthic infections, the worm loads are overdispersed in the host population with a small proportion of the available hosts harboring the majority of the worms. The genetic analyses of the worm burden data were designed to determine if the variation in levels of infection could be attributed to genetic factors.

As with our studies of *T. cruzi* infection, the first step in our analysis of worm infections in the Jirel population was to determine if the observed variation in worm loads could be attributable to genetic factors. A variance component approach was utilized to determine how much of the variation in each of the measures of worm burden (egg count, worm count, and worm weight) was due to genetic effects.

It was determined that there was significant evidence of genetic components influencing susceptibility to infection with *Ascaris*. Between 30 and 40 per cent of the variation in each of the traits could be attributed to genetic factors. The heritability of eggs per gram of faeces in the first year of this longitudinal study was 0.291 ($p < 0.0001$) (Williams-Blangero *et al.*, 1999). Similarly the heritability of worm count in the first year of the study was 0.359 ($p < 0.0001$), and the heritability of worm weight in the first year was 0.335 ($p < 0.0001$) (Williams-Blangero *et al.*, 1999). The heritabilities for worm burdens are remarkably similar across the years of the study (Williams-Blangero *et al.*, 1999; Williams-Blangero, unpublished data).

It was found that common household effects accounted for a small but significant amount of the variation in worm loads. Environmental effects explained between 3 and 13 per cent of the total variance in worm burden as assessed by egg counts, worm counts, and total worm weight (Williams-Blangero *et al.*, 1999).

The strong evidence for genetic determination of susceptibility to *Ascaris* infection led to development of a genome scan for the specific individual genes influencing susceptibility to this parasitic infection. The first stage of the genome scan in 444 members of a complex and genetically informative branch of the Jirel pedigree was recently completed, utilizing eggs per gram of faeces as the measure of worm burden (Williams-Blangero *et al.*, 2002b). For each individual 375 genetic markers were typed, and variance component linkage analysis, as implemented in the computer program SOLAR (Almasy and Blangero, 1998), was used to determine if the worm burden trait co-segregated with genetic markers in particular chromosomal regions.

The results of the genome scan provided the first evidence of specific genetic loci influencing variation in susceptibility to roundworm infection (Williams-Blangero *et al.*, 2002b). Genes located on chromosomes 1 and 13 exerted significant effects on *Ascaris* worm loads as measured by eggs per gram of faeces. In addition to the significant linkage signals for these two genes, there was a suggestive linkage to a region on chromosome 8 (Williams-Blangero *et al.*, 2002b).

The signal on chromosome 13 is near a candidate gene, *TNSF13B* (Williams-Blangero *et al.*, 2002b). This gene is a member of the tumor necrosis factor super family and regulates B cell activation and Ig secretion (Moore *et al.*, 1999; Schneider *et al.*, 1999; Yan *et al.*, 2000). While there is no clear candidate at or near the linkage signal on chromosome 1, the signal on chromosome 8 is near *IL7* (Williams-Blangero *et al.*, 2002b). This gene's product is found in intestinal epithelial cells and has been shown to deter development of helminths (Watanabe *et al.*, 1995; Wolowczuk *et al.*, 1999). When our genome scan is completed it may reveal additional genes with significant detectable effects on roundworm infection.

Knowledge of the individual genes affecting susceptibility to *Ascaris* infection will ultimately facilitate characterization of the physiological mechanisms involved. Identification of new mechanisms may provide novel targets for the design of improved anthelminthic drugs or of chemopreventions. In addition, further characterization of the allelic variants associated with susceptibility may provide a means of targeting treatments to those individuals likely to have high worm burdens and hence most likely to experience adverse consequences of infection.

Hookworm infection as an example of a parasitic disease with differential response to infection

Worldwide, 1298 million people are infected with hookworm (Crompton, 1999; Stephenson, 2002). Hookworm infection is one of the more serious of the soil-transmitted helminthiases because of its association with blood loss from the intestine. The worms range in size from a maximum of 10 mm for male worms to a maximum of 15 mm for female worms (Cheesbrough, 1987; Hotez, 1999). The volume of blood loss is a function of the number of adult worms in the gut, with a loss of between 0.03 to 0.15 ml per worm per day (Stephenson, 2002). As a result, heavy hookworm infections can lead to serious haematological consequences including anaemia (Pritchard *et al.*, 1991; Brooker *et al.*, 1999; Hotez, 1999). The disease diminishes the body's iron stores, and can be particularly severe in areas where the local diet has a low iron content (Pritchard *et al.*, 1991; Bulto *et al.*,

1992). Individuals infected with hookworm may have a significantly reduced work capacity (WHO, 1987; Wilson *et al.*, 1999; Gilgen *et al.*, 2001).

Hookworm infection is a disease for which differential response to infection may be observed. For example, there can be variation in the level of anaemia among individuals with similar worm loads. It was assessed whether or not the genetic determinants of haemoglobin levels vary between infected and uninfected individuals in the Jirel population participating in the Jiri Helminth Project described earlier.

The sample for this analysis consisted of 1199 members of the single Jirel pedigree for whom hookworm data were available. The measure of worm load was eggs per gram of faeces. A simple quantitative genetic approach was utilized to determine how much of the variation in haemoglobin was attributable to genetic factors in infected individuals, uninfected individuals, and in the population as a whole. The variance decomposition approach was used and the total variance in the trait was decomposed into components attributable to additive genetic variance (h^2), common environmental variance, and random environmental variance.

Hookworm infections are prevalent in the Jirel population, with 57 per cent of the 1199 individuals sampled being infected as determined by the Kato–Katz thick smear technique. Worm burden data were available for a subset of 1007 individuals from whom 96-hour faecal samples were collected following treatment with albendazole (400 mg). The average worm load in sampled individuals was 11.3 worms with a maximum of 300 worms expelled. Mixed hookworm infections were common. Fifty-seven per cent of infected individuals expelled both *Necator americanus* and *Ancylostoma duodenale*, 34.5 per cent expelled only *N. americanus*, and 8.5 per cent expelled only *A. duodenale*. The average haemoglobin level in the 1007 sampled individuals was 13.57 with a standard deviation of 1.59.

In this population, the estimated heritability of haemoglobin concentration is 0.399 ± 0.052, indicating that approximately 40 per cent of the variation in haemoglobin in the population as a whole is attributable to genetic factors. However, the heritability of haemoglobin levels in uninfected individuals is 0.54, indicating that over 50 per cent of the variation in haemoglobin levels among individuals not infected with hookworm is due to genetic factors. By contrast, the heritability of haemoglobin levels in individuals harboring hookworm is 0.38. If you consider infection status as a biological environment, the contrasting genetic variances provide evidence of genotype by environment interaction in the genetic control of haemoglobin; that is, the genetic influences on haemoglobin levels differ when an individual is infected with hookworm.

The genetic correlation between haemoglobin levels in infected and uninfected individuals is both significantly different form zero ($p = 0.001$) and significantly different from one ($p = 0.031$) indicating that while some common genetic factors influence haemoglobin in both infected and uninfected individuals, other genetic factors are not shared between the two states. A genetic correlation of 1.0 is expected in the absence of genotype-by-environment interaction (Blangero, 1993). The genetic correlation indicates that there may be a genetic basis to the response of haemoglobin levels to hookworm burden.

The evidence of genotype-by-infection status interaction in the determination of haemoglobin levels in this population suggests that joint linkage analysis of haemoglobin levels and hookworm burden may aid us in the search for the individual genes that influence the pathological outcomes associated with hookworm infection.

Conclusions

The examples presented in this chapter demonstrate that genetic epidemiological approaches can yield new information which will improve understanding of the parasitic diseases that persist as public health problems because they have no cure and/or because individuals may be readily reinfected. In addition, genetic studies may be informative for understanding variability in pathological response to infection. The results of such studies, and particularly of genome scans, will suggest new mechanisms to target in the development of improved chemotherapeutic agents. In addition, characterization of the specific genes influencing susceptibility to parasitic disease could ultimately provide a mechanism by which available drug treatments could be targeted to those most likely to experience severe disease outcomes.

Acknowledgements

We thank the participants of the Posse Family Health Study and the Jiri Helminth Project for their generous cooperation. The data collection and informed consent process in Brazil were approved by the Institutional Review Board (IRB) at the University of Texas in San Antonio, the Ethical Committee of the University of Brasilia, and the Brazilian Ministry of Health. The Posse Family Health Study is currently approved by the IRB at the University of Texas, San Antonio, the Ethical Committee at FIOCRUZ, Belo Horizonte, Brazil, and the Brazilian Ministry of Health. The data collection and informed consent process in Nepal are approved by the IRB at the University of Texas in San Antonio and the Nepal Health Research Council in Kathmandu, Nepal. This research was supported by NIH grants HL66480, AI44406, and AI37091 to S. Williams-Blangero. The new statistical methods for analysis of infectious disease data and the computer program SOLAR were developed under NIH grant MH59490 to John Blangero. SOLAR is available through the Southwest Foundation for Biomedical Research website at www.sfbr.org.

References

Abel, L., Demenais, F., Prata, A., Souza, A.E. and Dessein, A. (1991) 'Evidence for the segregation of a major gene in human susceptibility/resistance to infection by *Schistosoma mansoni*', *Am J Hum Genet* 48: 959–70.

Abel, L., Cot, M., Mulder, L., Carnevale, P. and Feingold, J. (1992) 'Segregation analysis detects a major gene controlling blood infection levels in human malaria', *Am J Hum Genet* 50: 1308–17.

Almasy, L.A. and Blangero, J. (1998) 'Multipoint quantitative trait linkage analysis in general pedigrees', *Am J Hum Genet* 62: 1198–211.

Beard, C., Young, D., Butler, J. and Evan, D. (1988) 'First isolation of *Trypanosoma cruzi* from a wild caught *Triatoma sanguisuga* (Le Conte) (Hemiptera: Triatominae) in Florida', *J Parasitol* 74: 343–4.

Blangero, J. (1993) 'Statistical genetic approaches to human adaptability', *Hum Biol* 65: 941–66.

Blangero, J., Williams, J.T. and Almasy, L. (2000) 'Quantitative trait locus mapping using human pedigrees', *Hum Biol* 72: 35–62.

Bradley, K.K., Bergman, D.K., Woods, J.P., Crutcher, J.M. and Kirchhoff, L.V. (2000) 'Prevalence of American trypanosomiasis (Chagas disease) among dogs in Oklahoma', *J Am Vet Med Assoc* 217: 1853–7.

Brooker, S., Peshu, N., Warn, P.A., Mosobo, M., Guyatt, H. and Marsh, K. (1999) 'The epidemiology of hookworm infection and its contribution to anaemia among pre-school children on the Kenya coast', *Trans Royal Soc Trop Med Hyg* 93: 240–6.

Bulto, T., Meskal, F.H., Endeshaw, T. and Dejene, A. (1992) 'Prevalence of hookworm infection and its associations with low haematocrit among resettlers in Gambela, Ethiopia', *Trans Roy Soc Trop Med Hyg* 86: 706–18.

Chai, J.-Y., Seo, B.-S. and Lee, S.-H. (1983) 'Epidemiological studies on *Ascaris lumbricoides* reinfection in rural communities in Korea. II. Age-specific reinfection rates and familial aggregation of the reinfected cases', *Korean J Parasitol* 21: 142–9.

Chan, M.-S. (1997) 'The global burden of intestinal nematode infections – fifty years on', *Parasitol Today* 13: 438–43.

Chan, M.-S., Medley, G.M., Jamison, D. and Bundy, D.A.P. (1994) 'The evaluation of potential global morbidity attributable to intestinal nematode infections', *Parasitology* 109: 373–387.

Cheesbrough, M. (1987) *Medical Laboratory Manual for Tropical Countries*, Oxford: Butterworth-Heinemann.

Chrungroo, R.K., Hangloo, V.K., Faroqui, M.M. and Khan, M. (1992) 'Surgical manifestations and management of ascariasis in Kashmir', *J Ind Med Assoc* 90:171–4.

Colorado, I.A., Acquatella, F., Catalioti, F., Fernandez, M.T. and Layrisse, Z. (2000) 'HLA class II DRB1, DPB1 polymorphisms and cardiomyopathy due to *Trypanosoma cruzi* chronic infection', *Hum Immunol* 61: 320–5.

Crompton, D.W.T. (1992) 'Ascariasis and childhood malnutrition', *Trans R Soc Trop Med Hyg* 86: 577–9.

Crompton, D.W.T. (1999) 'How much human helminthiasis is there in the world?', *J Parasitol* 85: 397–403.

Di Pentima, M.C., Hwang, L.Y., Skeeter, C.M. and Edwards, M.S. (1999) 'Prevalence of antibody to *Trypanosoma cruzi* in pregnant Hispanic women in Houston', *Clin Infec Dis* 28: 1281–5.

Dyer, T.D., Blangero, J., Williams, J.T., Goring, H.H. and Mahaney, M.C. (2001) 'The effect of pedigree complexity on quantitative trait linkage analysis', *Genet Epidemiol* 21: S236–S243.

Dyke, B. (1989) *PEDSYS. A Pedigree Data Management System. Users Manual*, PGL Technical Report No. 2, San Antonio, TX: Southwest Foundation for Biomedical Research.

Dykes, C.W. (1996) 'Genes, disease and medicine', *Br J Clin Pharmacol* 42: 683–95.

Fairlamb, A.H. (1999) 'Future prospects for the chemotherapy of Chagas' disease', *Medicina (B. Aires)* 59: 179–87.

Fauci, A.S. (2001) 'Infectious diseases: considerations for the 21st century', *Clin Infec Dis* 32: 675–85.

Feitosa, M.F. and Krieger, H. (1993) 'A model for intra-familial distribution of an infectious disease (Chagas' disease)', *Mem Inst Oswaldo Cruz* 88: 231–3.

Fernandez-Mestre, M.T., Layrisse, Z., Montagnani, S., Acquatella, H., Catalioti, F., Matos, M., Balbas, O., Makhatadze, N., Dominguez, E., Herrera, F. and Madrigal, A. (1998) 'Influence of the HLA class II polymorphism in chronic Chagas' disease', *Parasite Immunol* 20: 197–203.

Forrester, J.E., Scott, M.E., Bundy, D.A.P. and Golden, M.H.N. (1988) 'Clustering of *Ascaris lumbricoides* and *Trichuris trichiura* infections within households', *Trans Royal Soc Trop Med Hyg* 82: 282–8.

Garcia, A., Abel, L., Cot, M., Richard, P., Ranque, S., Feingold, J., Demenais, F., Boussinesq, M. and Chippaux, J.P. (1999) 'Genetic epidemiology of host predisposition to microfilaremia in human loiasis', *Trop Med Int Health* 4: 565–74.

Garcia Zapata, M.T.A. and Marsden, P.D. (1986) 'Chagas' disease', *Clin Trop Med Commun Dis* 1: 557–85.

Gelbert, L.M. and Gregg, R.E. (1997) 'Will genetics really revolutionize the drug discovery process?', *Curr Opin Biotechnol* 8: 669–74.

Gilgen, D., Mascie-Taylor, C.G.N. and Rosetta, L. (2001) 'Intestinal helminth infections, anaemia and labour productivity of female tea pluckers in Bangladesh', *Trop Med Int Health* 6: 49–57.

Gupta, M.C. (1990) 'Effects of ascariasis upon nutritional status of children', *J Trop Pediatrics*, 36: 189–91.

Gurtler, R.E., Cecere, M.C., Rubel, D.N. and Schweigmann, N.J. (1992) 'Determinants of the domiciliary density of *Triatoma infestans*, vector of Chagas disease', *Med Vet Entomol* 6: 75–83.

Holland, C.V. and Kennedy, M.W. (2002) 'Preface' in C.V. Holland and M.W. Kennedy (eds), *The Geohelminths: Ascaris, Trichuris, and hookworm*, Boston: Kluwer, pp. 11–14.

Holland, C.V., Crompton, D.W., Asaolu, S.O., Crichton, W.B., Torimiro, S.E., and Walters, D.E. (1992) 'A possible genetic factor influencing protection from infection with *Ascaris lumbricoides* in Nigerian children', *J Parasitol* 78: 915–16.

Hopper, J.L. and Mathews, J.D. (1982) 'Extensions to multivariate normal models for pedigree analysis', *Ann Hum Genet* 46: 373–383.

Hotez, P.J. (1999) 'Hookworm infections', in R.L. Guerrant, D.H. Walker and R.F. Weller. (eds) *Tropical Infectious Diseases: Principles, Pathogens, and Practice*, New York: Churchill Livingstone, pp. 966–74.

Ianelli, C.J., Edson, C.M. and Thorley-Lawson, D.A. (1997) 'A ligand for human CD48 on epithelial cells', *J Immunol* 159: 3910–20.

Kinnamon, K.E., Poon, B.T., Hanson, W.L. and Waits, V.B. (1997) 'In pursuit of drugs for American trypanosomiasis: evaluation of some "standards" in the mouse model', *Proc Soc Exp Biol Med* 216: 424–8.

Kirchhoff, L.V. (1989) 'Is *Trypanosoma cruzi* a new threat to our blood supply?', *Ann Intern Med* 111: 773–5.

Kirchhoff, L.V. (1993a) 'American trypanosomiasis (Chagas' disease) – a tropical disease now in the United States', *New Engl J Med* 329: 639–44.

Kirchhoff, L.V. (1993b) 'Perspectives on Chagas' disease in Latin America and the United States', in D.H. Walker (ed.), *Global Infectious Diseases: Prevention, Control, and Eradication*, New York: Springer Verlag, pp. 219–26.

Kirchhoff, L.V. (1999) 'American trypanosomiasis (Chagas' disease)', in R.L. Guerrant, D.H. Walker. and R.F. Weller (eds), *Tropical Infectious Diseases: Principles, Pathogens, and Practice*, New York: Churchill Livingstone, pp. 785–96.

Kirchhoff, L.V. and Neva, F.A. (1985) 'Chagas' disease in Latin American immigrants', *J Amer Med Assoc* 254: 3058–60.

Klyushnenkova, E.N., Li, L., Armitage, R.J. and Choi, Y.S. (1996) 'CD48 delivers an accessory signal for CD40-mediated activation of human B cells', *Cell Immunol* 174: 90–8.

Lange, K. and Boehnke, M. (1983) 'Extensions to pedigree analysis. IV. Covariance components models for multivariate traits', *Am J Med Genet* 14: 513–24.

Lange, K., Weeks, D. and Boehnke, M. (1988) 'Programs for pedigree analysis: MENDEL, FISHER, and dGENE', *Genet Epidemiol* 5: 4712.

Llop, E.R., Rothhammer, F., Acuña, M., Apt, W. and Arribada, A. (1991) 'Antigenos HLA en cariopatas chagasicos: Nueva evidencia basada en un analisis de casos y controles', *Rev Med Chile* 119: 633–6.

MacCluer, J.W., Stern, M.P., Almasy, L., Atwood, L.A., Blangero, J., Comuzzie, A.G., Dyke, B., Haffner, S.M., Henkel, R.D., Hixson, J.E., Kammerer, C.M., Mahaney, M.C., Mitchell, B.D., Rainwater, D.L., Samollow, P.B., Sharp, R.M., VandeBerg, J.L. and Williams, J.T. (1999) 'Genetics of atherosclerosis risk factors in Mexican Americans', *Nutr Rev* 57: S59–S65.

Marquet, S., Abel, L., Hillaire, D., Dessein, H., Kalil, J., Feingold, J., Weissenbach, J. and Dessain, A.J. (1996) 'Genetic localization of a locus controlling intensity of infection by *Schistosoma mansoni* on chromosome 5q31–q33', *Nat Genet* 14: 181–184.

Mitchell, B.D., Kammerer, C.M., Blangero, J., Mahaney, M.C., Rainwater, D.L., Dyke, B., Hixson, J.E., Henkel, R.D., Sharp, R.M., Comuzzie, A.G., VandeBerg, J.L., Stern, M.P. and MacCluer, J.W. (1996) 'Genetic and environmental contributions to cardio-vascular risk factors in Mexican Americans. The San Antonio Family Heart Study', *Circulation* 94: 2159–70.

Moore, P.A., Belvedere, O., Orr, A., Pieri, K., LaFleur, D.W., Feng, P., Soppet, D., Charters, M., Gentz, R., Parmelee, D., Li, Y., Galperina, O., Giri, J., Roschke, V., Nardelli, B., Carrell, J., Sosnovtseva, S., Greenfield, W., Ruben, S.M., Olsen, H.S., Fikes, J. and Hilbert, D.M. (1999) 'BlyS: member of the tumor necrosis factor family and B lympho-cyte stimulator', *Science* 285: 260–3.

Morris, S.A., Tanowitz, H.B., Wittner, M. and Bilezikian, J.P. (1990) 'Pathophysiological insights into the caridomyopathy of Chagas' disease', *Circulation* 82: 1900–9.

Murray, C.J.L. (1994) 'Quantifying the burden of disease: the technical basis for disability-adjusted life years', in C.J.L. Murray and A.D. Lopez (eds) *Global Comparative Assessments in the Health Sector.* Geneva: World Health Organization.

Navin, T.R., Roberto, R.R., Juranek, D.D., Limpakarnjanarat, K., Mortenson, E.W., Clover, J.R., Yescott, R.E., Taclindo, C., Sieurer, F. and Allain, D. (1985) 'Human and sylvatic *Trypanosoma cruzi* infection in California', *Am J Pub Health* 75: 366–369.

Nickerson, P., Orr, P., Schroeder, M., Sekla, L. and Johnston, J. (1989) 'Transfusion associated *Trypanosoma cruzi* infection in a non-endemic area', *Ann Intern Med* 111: 851–3.

Nieto, A., Beraun, Y., Collado, M.D., Caballero, A., Alonso, A., Gonzalez, A. and Martin, J. (2000) 'HLA haplotypes are associated with differential susceptibility to *Trypanosoma cruzi* infection', *Tissue Antigens* 55: 195–8.

Nogueira, N. and Coura, J. (1990) 'American trypanosomiasis (Chagas' disease)', in K.S. Warren and A.A.F. Mahmoud (eds), *Tropical and Geographic Medicine*, New York, McGraw-Hill, pp. 281–96.

Ochs, D.E., Hnilica, V.S., Moser, D.R., Smith, J.H. and Kirchhoff, L.V. (1996) 'Postmortem diagnosis of autochthonous acute chagasic myocarditis by polymerase chain reaction amplification of species-specific DNA sequence of *Trypanosoma cruzi*', *Am J Trop Med Hyg* 64: 526–9.

PAHO (1982) 'Chagas' disease', *Epidemiol Bull Pan Amer Health Org* 3: 1–4.

Pritchard, D.I., Quinnell, R.J., Moustafa, M., McKean, P.G., Slater, A.F.G., Raiko, A., Dale, D.D.S. and Keymer, A.E. (1991) 'Hookworm (*Necator americanus*) infection and storage iron depletion', *Trans Roy Soc Trop Med Hyg* 85: 235–8.

Ramsay, C.E., Hayden, C.M., Tiller, K.J., Burton, P.R., Hagel, I., Palenque, M., Lynch, N.R., Goldblatt, J. and LeSouef, P.N. (1999) 'Association of polymorphisms in the beta2-adrenoreceptor with higher levels of parasitic infection', *Hum Genet* 104: 269–74.

Schmunis, G.A. (1991) '*Trypanosoma cruzi*, the etiologic agent of Chagas' disease: status in the blood supply in endemic and nonendemic countries', *Transfusion* 31: 547–57.

Schneider, P., MacKay, F., Steiner, V., Hofmann, K., Bodmer, J.L., Holler, N., Ambrose, C., Lawton, P., Bixler, S., Acha-Orbea, H., Valmori, D., Romero, P., Werner-Favre, C., Zubler, R.H., Browning, J.L. and Tschopp, J. (1999) *J Exp Med* 189: 1747–56.

Schofield, C.J. and Dias, J.C.P. (1999) 'The Southern Cone initiative against Chagas' disease', *Adv Parasitol* 42: 1–27.

Seltzer, E. and Barry, M. (1999) 'Ascariasis', in R.L. Guerrant, D.H. Walker and R.F. Weller (eds), *Tropical Infectious Diseases: Principles, Pathogens, and Practice*, New York: Churchill Livingstone, pp. 959–65.

Silva, N.R. de, Chan, M.S. and Bundy, D.A.P. (1997a) 'Morbidity and mortality due to ascariasis: re-estimation and sensitivity analysis of global numbers at risk', *Trop Med Int Health* 2: 519–28.

Silva, N.R. de, Guyatt, H. and Bundy, D.A. (1997b) 'Worm burden in intestinal obstruction caused by *Ascaris lumbricoides*', *Trop Med Int Health* 2: 189–90.

Silva, N.R. de, Guyatt, H. and Bundy, D.A. (1997c) 'Anthelmintics: a comparative review of their clinical pharmacology', *Drugs* 53: 769–788.

Stephenson, L.S. (1987) *Impact of Helminth Infections on Human Nutrition*, London: Taylor and Francis.

Stephenson, L.S. (2002) '*Pathophysiology of intestinal nematodes*', in C.V. Holland and M.W. Kennedy (eds), *The Geohelminths: Ascaris, Trichuris, and hookworm*, Boston: Kluwer, pp. 39–63.

Stephenson, L.S., Latham, M.C. and Ottesen, E.A. (2000) 'Malnutrition and parasitic helminthic infections', *Parasitology* 121: 523–38.

Stoppani, A.O. (1999) 'The chemotherapy of Chagas' disease', *Medicine (B. Aires)* 59 (Suppl. 2): 147–65.

Tanowitz, H.B., Morris, S.A., Factor, S.M., Weiss, L.M., and Wittner, M. (1992) 'Parasitic disease of the heart I. Acute and chronic Chagas' disease: trends in immunological research and prospects for immunoprophylaxis', *Bull World Health Org* 57: 697–710.

Tanowitz, H.B., Weiss, L.M. and Wittner, M. (1994) 'Diagnosis and treatment of common intestinal helminths. II: Common intestinal nematodes', *Gastroenterologist* 2: 39–49.

Teixeira, A.R.L. (1979) 'Chagas' disease: trends in immunological research and prospects for immunoprophylaxis', *Bull World Health Org* 57: 697–710.

Teixeira, A.R.L., Cordova, J.C., Souto Maior, I. and Solorzano, E. (1990a) 'Chagas' disease: lymphoma growth in rabbits treated with benzinidazole', *Am J Trop Med Hyg* 43: 146–58.

Teixeira, A.R.L., Silva, R., Neto, E.C., Santano, J.M. and Rizzo, L.V. (1990b) 'Malignant, non-Hodgkins lymphoma in *Trypanosoma cruzi* infected rabbits treated with nitroarenes', *J Comp Pathol* 103: 37–48.

Tissot, C., Rebouissou, C., Klein, B. and Mechti, N. (1997) 'Both human alpha/beta and gamma interferons upregulate the expression of CD48 cell surface molecules', *J Interferon Cytokine Res* 17: 17–26.

Towne, N., Blangero, J., Subedi, J., Rai, D.R., Upadhayay, R.P. and Williams-Blangero, S. (2000) 'Effects of heredity and helminthic infection on the growth of Nepali children', *Am J Trop Med Hyg* 62: 292–3.

Vassena, C.V., Picollo, M.I. and Zerba, E.N. (2000) 'Insecticide resistance in Brazilian *Triatoma infestans* and Venezuelan *Rhodnius prolixus*', *Med Vet Entomol* 14: 51–5.

Watanabe, M., Ueno, Y., Yajima, T., Iwao, Y., Tsuchiya, M., Ishikawa, H., Aiso, S., Hibi, T. and Ishii, H. (1995) 'Interleukin 7 is produced by human intestinal epithelial cells and regulates the proliferation of intestinal mucosal lymphocytes', *J Clin Invest* 95: 2945–53.

WHO (1987) 'Prevention and Control of intestinal parasitic infections', *WHO Tech Rep Ser* 749, Geneva: World Health Organization.

WHO (1991) 'Control of Chagas' disease', *WHO Tech Rep Ser* 811, Geneva: World Health Organization.

WHO (1993) *Chagas' Disease: Tropical Diseases, Tropical Disease Research Progress, 1991–1992.* Eleventh Programme Report of the UNDP/World Bank/WHO Special Programme for Research and Training in Tropical Diseases, Geneva: World Health Organization, pp. 67–75.

WHO (1999) 'Chagas' disease', *Weekly Epidemiological Record* 35: 290–2.

Williams, D., Burke, G. and Hendley, J.O. (1974) 'Ascariasis: a family disease', *J Pediatrics* 84: 853–4.

Williams-Blangero, S. (1990) 'Population structure of the Jirels: patterns of mate choice', *Am J Phys Anthropol* 82: 61–72.

Williams-Blangero, S. and Blangero, J. (2002) '*Human host susceptibility to helminthic infection*', in C.V. Holland and M.V. Kennedy (eds), *The Geohelminths: Ascaris, Trichuris, and Hookworm*, Boston: Kluwer, pp. 167–184.

Williams-Blangero, S., Blangero, J. and Bradley, M. (1997a) 'Quantitative genetic analysis of susceptibility to hookworm infection in a population from rural Zimbabwe', *Hum Biol* 69: 201–8.

Williams-Blangero, S., VandeBerg, J.L., Blangero, J. and Teixeira, A.R.L. (1997b) 'Genetic epidemiology of seropositivity for *Trypanosoma cruzi* infection in rural Goiás, Brazil', *Am J Trop Med Hyg* 57: 538–43.

Williams-Blangero, S., Subedi, J., Upadhayay, R.P., Manral, D.P., Rai, D.R., Jha, B., Robinson, E.S. and Blangero, J. (1999) 'Genetic analysis of susceptibility to infection with *Ascaris lumbricoides*', *Am J Trop Med Hyg* 60: 921–6.

Williams-Blangero, S., McGarvey, S.T., Subedi, J., Wiest, P.M., Upadhayay, R.P., Rai, D.R., Jha, B., Olds, G.R., Guanling, W. and Blangero, J. (2002a) 'Genetic component to susceptibility to *Trichuris trichiura*: evidence from two Asian populations', *Genet Epidemiol* 22: 254–64.

Williams-Blangero, S., VandeBerg, J.L., Subedi, J., Aivaliotis, M.J., Rai, D.R., Upadhayay, R.P., Jha, B. and Blangero, J. (2002b) 'Genes on chromosomes 1 and 13 have significant effects on *Ascaris* infection', *Proc Natl Acad Sci*, USA 99: 5533–8.

Wilson, W.M., Dufour, D.L., Staten, L.K., Barac-Nieto, M., Reina, J.C. and Spurr, G.B. (1999) 'Gastrointestinal parasitic infection, anthropometrics, nutritional status, and physical work capacity in Colombian boys', *Am J Hum Biol* 11: 763–71.

Wolowczuk, I., Nutten, S., Roye, O., Delacre, M., Capron, M., Murray, R.M., Trottein, F. and Auriault, C. (1999) 'Infection of mice lacking interleukin-7 (IL-7) reveals an unexpected role for IL-7 in the development of the parasite *Schistosoma mansoni*', *Infec Immunol* 67: 4183–90.

Yabsley, M.J. and Noblet, G.P. (2002) 'Serpoprevalence of *Trypanosoma cruzi* in raccoons from South Carolina and Georgia', *J Wildl Dis* 38: 75–83.

Yaeger, R.G. (1988) 'The prevalence of *Trypanosoma cruzi* infection in armadillos collected at a site near New Orleans, Louisiana', *Am J Trop Med Hyg* 38: 323–6.

Yan, M., Marsters, S.A., Grewal, I.S., Wang, H., Ashkenazi, A. and Dixit, V.M. (2000) 'Identification of a receptor for BlyS demonstrates a crucial role in humoral immunity', *Nat Immunol* 1: 37–41.

Zayas, C.F., Perlino, C., Caliendo, A., Jackson, D., Martinez, E.J., Tso, P., Heffron, T.G., Logan, J.L., Herwaldt, H.L., Moore, A.C., Steurer, F.J., Bern, C. and Maguire, J.H. (2002) 'Chagas disease after organ transplantation – United States, 2001', *Morbidity and Mortality Weekly Report* 51: 210–12.

Zerba, E.N. (1999) 'Susceptibility and resistance to insecticides of Chagas disease vectors', *Medicina (B Aires)* 59: 41–6.

Zicker, F., Smith, P.G., Netto, J.C.A., Oliveira, R.M. and Zicker, E.M.S. (1990) 'Physical activity, opportunity for reinfection, and sibling history of heart disease as risk factors for Chagas' cardiopathy', *Am J Trop Med Hyg* 43: 498–505.

5 Urban pollution, disease and the health of children

L.M. Schell and Elaine A. Hills

Urban growth

Since human populations established sedentary life approximately 10,000 years ago, urban population growth has been continuous, and over the past two hundred years, urban growth has been dramatic. By 2030 more than 60 per cent of the world population will be living in urban places (Department of Economic and Social Affairs 2000). In the more economically developed regions, 84 per cent of the population will be urban and in the lesser developed areas, 57 per cent of the population will be urban (Department of Economic and Social Affairs 2000).

A substantial slowdown in urban growth is not likely despite slight changes in the rate of urbanization in recent years. Although the rate of urban growth declines, the base of urban population is large and increasing, and the average annual increment in numbers of persons is steadily becoming larger. Between 1990 and 1995, a period with a relatively low rate of urban population growth, 59 million new urban dwellers were added to the world's population, 98 per cent of whom were in less developed countries. The fastest and greatest urban growth will be in the lesser and least developed countries, areas that anthropologists often study. Indeed, the less developed countries will contain 80 per cent of the world's urban population by 2030 (Department of Economic and Social Affairs 2000).

A less visible but equally important trend is the growth of mega-cities, cities with a population of 10 million or more. In 1950 only New York was a mega-city. Ten years later Tokyo had become one and by 1975 there were five. In 1995 there were 14. By 2015 there are expected to be 26 worldwide and 22 of these will be in the less developed countries.

There is no apparent barrier to continued urban growth. Infectious disease might be considered a barrier to further growth, but in the past and in many places currently, urban populations have grown despite infectious disease. The growth engine is fueled by the economic advantages that individuals experience or expect to experience as a result of their urban residence. This motivation is similar to the pull of urban places that existed in 18th- and 19th-century Europe but, unlike earlier times, it is not balanced by a high urban death rate to keep the urban population from growing rapidly (Weber 1967; Bogin 1988).

Not only are urban places growing but the effect of urban places on non-urban ones is growing also. In the US and many other countries, the commodification and advertising of diet, clothing, information/entertainment and leisure activities have led to a blurring of the boundary between urban and non-urban places. The once-isolated rural residents now consume many of the same products as urbanites and perform many of the same activities.

Human biology of urbanism and urbanization

Urban ecology and lifestyle is characterized by features that differ from the selective forces that shaped our species through millennia of evolution before urbanization began. As such, urban ecology and lifestyle pose biological challenges of interest to anthropologists. These challenges relate to reproduction, human growth and aging, and the demographic and genetic parameters of morbidity and mortality, all areas anthropologists have studied to understand human evolution and biologic variation (Stinson *et al.* 2000).

Environmental features that affect any of these deserve attention from human biologists. There are five main components of urban ecology that are relevant to these research areas. They include psychosocial stress, alterations in diet and activity patterns with consequences for energy balance, steep social gradients, close juxtaposition of social groups relating most closely but not exclusively to increased transmission of infectious disease and opportunity for faster evolution of infectious disease, and increased pollutant exposure and burden (Schell and Ulijaszek 1999).

Here the focus will be on the effect of industrial pollution on human biology. Although organic pollution, i.e. waste from animals, humans, and some forms of commerce, has always been associated with urbanism, global contamination from industrial pollution is relatively new, and its consequences for human biology and health are only now being understood.

Pollution from industrial sources takes diverse forms and evidence for its effects on parameters of importance to human biologists in the study of evolution, adaptation and variation is so extensive that a comprehensive review is impossible here. Instead, two subjects are considered: 1) the effect of urban air pollution on mortality and child growth among urban residents, and 2) the impact of urban sources of pollution on people living in non-urban places.

The question of urbanism and health is approached by examining specific components of urbanism rather than by comparing health indices of urban and rural populations. The latter approach is not sufficiently specific to identify biological challenges of urban ecology and would gloss over the reasons for dissimilarity or similarity.

Recently the US Centers for Disease Control and Prevention compared health indices of urban and non-urban places (Eberhardt *et al.* 2001). With some exceptions, urban places, particularly urban fringe areas (suburbs), actually had better health indices than non-urban ones (Figure 5.1). For persons 1–24 years of age, the least urban area had the highest mortality rate. For persons 25–64 years of age, the central areas of large cities in metropolitan counties had the highest mortality rate, while the least urbanized areas had the next highest.

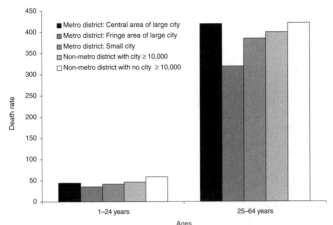

Figure 5.1 Mortality by level of urbanism in the US according to the CDC chartbook on urban and rural health (Eberhardt *et al.* 2001).

The pattern of health services provision in terms of physicians per person (Figure 5.2) partly explains these mortality statistics. The fact that occupational accidents are common causes of death in rural areas should also be considered. Differences in mortality rates by degree of urbanization are heavily influenced by socio-economic factors and occupational hazards as well as the distribution of health care facilities and services. Any summary statistic of urban–rural differences combines beneficial and detrimental influences and cannot be taken as an indicator of the healthfulness of any specific influence. It may be true that urban populations are healthier in some respects, at least presently in the US, but it is not true that all features of the urban ecology contribute to that healthfulness, and many are suspected of causing disease.

Pollution

Pollution is usually defined as an unwanted material or energy that is considered a threat to well-being, and it may be from natural sources or human production. Air pollution is a heterogeneous category insofar as it includes gases such as NO_x (oxides of nitrogen), SO_x (oxides of sulphur), CO (carbon monoxide), O_3 (ozone), as well as particulate matter (PM). Particulate matter is referred to in various ways, based upon the monitoring equipment with which the particulates were measured. Older monitoring equipment did not have the ability to precisely differentiate particles by size. Two categories of PM that reflect the use of older monitoring equipment include total suspended particulates (TSP) and suspended particulate matter (SPM), and include particles of any size floating in the air. With more refined measuring techniques, PM is more specifically represented. PM_{10}, or the coarse particle fraction, refers to particulate matter with an aerodynamic diameter less than or equal to 10 μm but greater than 2.5 μm. $PM_{2.5}$, also known as the fine

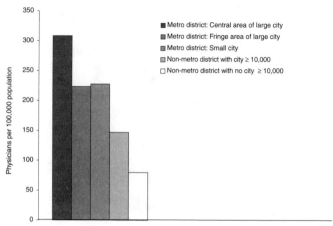

Figure 5.2 Number of physicians per capita by level of urbanism in the US according to the CDC chartbook on urban and rural health (Eberhardt *et al.* 2001).

particle fraction, refers to particulate matter with an aerodynamic diameter less than or equal to 2.5 mm.

Because techniques for measuring air pollution have improved so enormously over the last decade, earlier studies are difficult to compare with recent ones. Thus, this review focuses on air pollution studies published from 1995–2000 (the five years prior to the SSHB symposium on which this volume is based). We also limit our review to studies that analyze the effects of air pollution on the general population and exclude studies of vulnerable groups that may already display morbidity (e.g. asthma), predisposing them to the effects of air pollution (except in a few instances as noted). Accident-related mortality is disregarded since it is unlikely to be due to air pollution, except in rare instances (e.g. the London smog episodes of the early 1950s, see Waldbott (1978)). Finally, to take maximum advantage of advances in pollution measurement and monitoring as well as refined statistical methodologies, mortality was analyzed by type of air pollutant including O_3, CO, SO_x, PM, and NO_x. Mortality studies that do not analyze specific pollutants are not included in this review. Results for all-age mortality and pre-reproductive mortality were separated because pre-reproductive mortality is more closely related to reproductive fitness.

Air pollution and mortality

Though air pollutants have existed for centuries, the turning point in human health effects from exposure to air pollution is believed to have occurred with the Industrial Revolution, whence non-occupational exposure increased from the relatively low levels likely experienced prior to that period (Schell 1991a). Air pollution that resulted from industrialization was first recognized by dense smogs, as well as visible layers of smoke and soot in areas where industry flourished (United Nations Environment Program 1991). London, for example, became reputed for its deadly smog episodes during the late 19th and early 20th centuries, killing from 500 to 4000 individuals in each single episode. From study of these and other air pollution episodes that occurred through the mid-20th century, epidemiologists have generated a large body of literature from investigations of the health consequences of air pollutants.

Strong evidence for the effects of air pollution on health comes from studies of specific air pollutants and all-age mortality that have been conducted in rural and urban populations from Asia (Lee *et al.* 1999; Xu *et al.* 2000), North America (Schwartz 1991; Borja-Aburto *et al.* 1998; Moolgavkar 2000), and Europe (Zmirou *et al.* 1998; Peters *et al.* 2000). Our prior review of these studies (Schell and Hills 2002) demonstrated a remarkable consistency in the association of air pollution with mortality for monitored levels of particulates, SO_x, O_3, and CO but not NO_x.

Though not all studies provide such figures, the excess level of mortality that can be statistically attributed to air pollutants can be considerable. For instance, studies in the Teplice region of the Czech Republic (Peters *et al.* 2000) demonstrated that a 9.5 per cent increase in all-age mortality was associated with an increase of 100 $\mu g/m^3$ of TSP where the range of TSP levels during the study period was 17–940 $\mu g/m^3$. Similarly, in the Los Angeles region of the US, Moolgavkar (2000) demonstrated a significant association between SO_2 and total mortality with a 12.1 per cent change in daily mortality associated with a 10 ppb increase in levels of SO_2. The range of SO_2 in Los Angeles during this study period was 0–16 ppb, but in other regions of the US it has been documented to be twice as large.

These and other studies also provide information on the relative risk (RR) of death associated with air pollutants, with the risk often significant but only equating to a marginal increase (e.g. 1–10 per cent) in risk of mortality. At face value these RR figures, despite their statistical significance, do not appear very striking. However, the fact that these figures are being applied to substantial, often densely populated urban areas leads to an extremely high increase in the actual numbers of people that are at an increased risk of mortality when exposed to ambient air pollution.

Though human health effects often take a small role in current debates on greenhouse gas emissions, this is changing as a result of the strong and consistent evidence of the detrimental effects of air pollutants on human health. For instance, it has recently been determined that if greenhouse gas reduction measures were taken today, approximately 8 million deaths related to air pollution could be avoided worldwide by 2020 (Working Group on Public Health and Fossil-Fuel Combustion 1997), with the overwhelming majority of these deaths projected to occur in developing countries (see Figure 5.3). Dramatic as these figures are, they tell us little about a considerable, and evolutionarily significant portion of the world's population, the children.

It is only within the last ten years that studies have begun to focus on the effects of air pollution on pre-reproductive mortality, and there are far fewer studies on the subject than there are on all-age mortality. Those that exist focus on infants (children less than one year of age). Infants, unlike adults, are less likely to have moved from one pollutant exposure area to another during their lifetime (Pereira *et al.* 1998), making the strength of the association in infants particularly noteworthy.

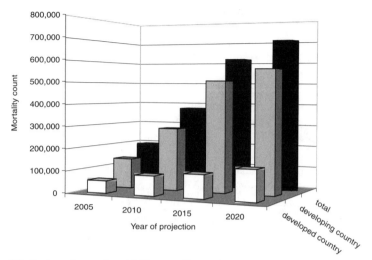

Figure 5.3   Projected annual avoidable mortality worldwide due to air pollution in developing and developed countries (Working Group on Public Health and Fossil-Fuel Combustion 1997).

Of the studies that have examined pre-reproductive mortality in relation to air pollution, most have found significant relationships between the two (Penna and Duchiade 1991; Bobak and Leon 1992, 1999a; Woodruff *et al.* 1997; Pereira *et al.* 1998; Loomis *et al.* 1999). Different studies focus on different stages at which the mortality occurred, with the phases being broken down into intrauterine, infant, neonatal, and postneonatal mortality. Only one study reviewed here did not demonstrate any relationship (Dolk *et al.* 2000) and this study may have lacked statistical power to detect an effect owing to a small sample size.

In contrast to the studies of all-age mortality, studies of sub-adult mortality demonstrate a consistent association with NO_x. This association was significant in varied populations in Mexico (Loomis *et al.* 1999), Brazil (Pereira *et al.* 1998), and the Czech Republic (Bobak and Leon 1992) and included intrauterine, infant, neonatal, and postneonatal stages. In Brazil (Pereira *et al.* 1998), a dose-dependent relationship between intrauterine death and NO_2 was found and this helps to establish a causal link rather than solely an associative one. Also noteworthy is the possibility of synergistic effects of pollutants. In Brazil (Pereira *et al.* 1998), an association between CO and intrauterine mortality was significant only in a combined model with other pollutants.

Studies on TSP and pre-reproductive mortality have been conducted in the Czech Republic in both an ecological study design (Bobak and Leon 1992) and a population-based case-control design (Bobak and Leon 1999a). The latter study design is unique and powerful in studies of air pollution exposure in humans, due to its use of individual-level information on exposure. Both studies yielded significant associations between TSP and pre-reproductive mortality, with the latter study only demonstrating this relationship for postneonatal respiratory mortality.

Postneonatal respiratory mortality was also significantly associated with SO_2 (Bobak and Leon 1992, 1999a, 1999b) in the Czech Republic. In Brazil (Pereira *et al.* 1998), SO_2 was associated with intrauterine mortality. These three cases of significant associations between pre-reproductive mortality and SO_2 were weak when considered in isolation but strong when considered in conjunction with other pollutants, again pointing to the clearer effects of pollutants when multi-pollutant exposure is considered.

Air pollution and prenatal growth

Within the studies of prenatal growth and air pollutants, the health end points measured include low birth weight (LBW), intrauterine growth retardation (IUGR), and prematurity.

In two different studies from the Czech Republic, significant associations between exposure to PM (including measures of TSP, PM_{10}, and $PM_{2.5}$) with LBW and prematurity were shown (Dejmek *et al.* 1999; Bobak 2000). IUGR was also associated with PM_{10} and $PM_{2.5}$ in one (Dejmek *et al.* 1999), but not both of these studies. TSP has been associated with prenatal growth outcomes in Beijing, China, where a significant relationship was demonstrated with the incidence of prematurity (Xu *et al.* 2000).

Carbon monoxide levels were monitored in both the Denver (Alderman *et al.* 1987) and Los Angeles (Ritz and Yu 1999) regions of the US, and assessed against the outcome of LBW. The Denver study did not show a significant effect on LBW, while the Los Angeles study did. However, the sample size for the study in Denver was 2870, far less than the sample size of 125,573 in the Los Angeles study. The Denver study may lack the statistical power to detect any relationship.

Studies monitoring SO_2 levels in relation to prenatal growth encompass populations in the Czech Republic, China, and the UK. Measures of prematurity were significantly increased in relation to SO_2 exposure in both the Czech Republic (Bobak 2000) and Beijing, China (Xu *et al.* 2000). Incidence of LBW was significantly associated with SO_2 levels in two different populations in the Czech Republic (Bobak and Leon 1999a; Bobak 2000), but not in areas surrounding cokeworks in the UK (Dolk *et al.* 2000). The measures in the UK did not single out SO_2, but rather measured SO_2 in a combined model with various pollutants. This may have altered the ability to detect an effect from SO_2, if one existed. Moreover, the authors suggest that they did not have the statistical power to demonstrate significant effects on health deriving from cokeworks, as they have been demonstrated for adults in the same regions.

Though most of the pollutants measured in relation to prenatal growth outcomes have shown significant associations, NO_2 does not. The one study reviewed here that monitored NO_2 levels (Bobak 2000) did so in relation to LBW, IUGR, and prematurity in the Czech Republic. In addition to being the only study to monitor NO_2 levels, it also was the only study reviewed here that showed no effect of air pollutants on the occurrence of prematurity. Conversely, this study agrees with all but one (Dejmek *et al.* 1999) of the studies reviewed here that show no relationship between IUGR and air pollutant exposure.

For a better understanding of the relationship between air pollutant levels and prenatal growth, future studies may benefit from more careful distinctions in their outcome variables. For instance, studies could indicate whether use of the outcome IUGR refers to proportionate or disproportionate growth retardation, though such a distinction may further impair the statistical power of studies. They also may benefit from selecting study areas with contrasting mixes of air pollutants.

Air pollution and postnatal growth

Recent studies of postnatal growth in relation to air pollution are rare (for an earlier review see Schell (1991a, 1991b)). Because of the paucity of recent studies, all recent reports are reviewed, including ones where specific pollutants were not measured separately.

Studies from Krakow, Poland (Jedrychowski *et al.* 1999), and Gelsenkirchen, Federal Republic of Germany (FRG) (Schlipköter *et al.* 1986), have linked postnatal growth to SO_2 and SPM levels. The Krakow study examined height velocity of nine-year-old children (Jedrychowski *et al.* 1999). The study in the FRG assessed bone maturation on children 7–9 years of age who were exposed to high levels of pollutants at both the initiation and the 10-year follow-up for the study. Both showed significantly delayed growth or maturation in relation to pollutant levels.

In a comparative study of more than ten thousand 7–12 year old children in polluted and non-polluted areas, a consistent effect was shown of reduced skeletal maturation in children residing in the polluted areas (Schmidt and Dolgner 1977). The authors point out that the effects may be confounded by factors other than the pollution levels, a common difficulty in interpreting comparative studies. One study found advanced maturation in relation to air pollution (Danker-Hopfe and Hulanicka 1995) that was attributed to past exposure to airborne lead. This raises the interesting possibility that pollutants not only depress growth but could alter it in other ways depending on the biochemical properties of the pollutant.

The ecological footprint of urbanism

The studies just reviewed have all taken the basic approach of examining the effect of an urban feature on the health of urban populations, rather than studying the effect of urban products on both urban and non-urban populations. However, many urban products have tremendous impact beyond the city borders and its statistical catchment area; air pollution is one example. This extensive reach of urban impact is really an application of the concept of the ecological footprint (Waskernagel and Rees 1996).

Urban places affect non-urban ones in several ways. Clearly air and water circulation patterns extend the influence of the city. Remote hunting and gathering communities of the North American Arctic have been affected by consumption of local animals contaminated with persistent organic pollutants that entered the ocean and atmosphere far away and have worked their way up the food chain and into large sea mammals that are preferred local foods (Johansen 2000). The city infrastructure requires electrical power that is usually generated in rural areas, and it also requires waste disposal which is done outside the city. Finally, in many instances manufacturing is located in rural areas. In general, urban pollution and the industrialization that supports urban populations have had effects on the global environment.

Pollution and endocrinology in non-urban areas

A seminal paper by Carlsen and colleagues reported a negative secular trend in measures of male reproductive function (Carlsen *et al.* 1992). A remarkable finding was that mean sperm density (count) had declined by approximately 50 per cent from 1940 to 1990 and seminal fluid volume had decreased significantly also. The authors concluded that the effect probably had environmental causes rather than a genetic one since the gene pool had not changed over the study period. They raised the possibility that testicular function could be affected by oestrogens or compounds with oestrogen-like activity derived from pollutants including pesticides, herbicides, and PCBs (polychlorinated biphenyls). The notion that chemicals could affect reproduction has been known since 1962 when Rachel Carson's book, *The Silent Spring*, was published, but the idea that human fertility might be affected seemed new. Reports following Carlsen's supported the initial findings while others did not confirm the idea of reduced human reproductive function in relation to certain environmental contaminants. In subsequent papers the rates of testicular cancer, the rate of hypospadius (a congenital malformation of the male reproductive tract), and the rate of cryptochordism (undescended testes) were all linked, with greater or lesser evidence, to environmental toxicants (Carlsen *et al.* 1995). The debate is now in high gear (Crisp *et al.* 1998; Bigsby *et al.* 1999; Bentley 2000), and concerns us because of anthropological interest in factors, such as male reproductive function, that affect evolution.

To investigate this and other functional outcomes related to endocrine disruption by environmental toxicants, a study was initiated to understand the impact of PCBs and several other toxicants that are common environmental contaminants on adolescent growth and maturation, including the endocrine parameters underlying both. Details of the study have been published before (Gallo *et al.* 2002; Schell *et al.* 2002; Schell and Tarbell 1998) and only a brief description is possible here.

The study community is the Mohawk Nation at Akwesasne. The community straddles the St. Lawrence River at the juncture of New York State and the Canadian provinces of Quebec and Ontario. In 1959 the St. Lawrence River became the St. Lawrence Seaway and ocean-going ships began to use it to travel to ports on the Great Lakes. Simultaneously, industries developed on the riverside and in only 30 years these have left a legacy of environmental contamination. On the Akwesasne community's western boundary are several industries and a toxic waste site where PCBs have entered the soil and the St. Lawrence river, thereby contaminating local fish. Fish have been a traditional food for the Akwesasne Mohawk community, and consumption of locally caught fish was considerable until the community was advised by local health departments in the early 1980s to limit consumption of local fish and game. The adolescents in the sample were born between 1978 and 1985, a period when people had already begun to heed the fish advisories issued by local health agencies and reduce their intakes of locally caught fish. However, women's consumption during their prepregnant years before the advisory notices were issued had given the mothers of the adolescents a toxicant burden.

Ingested PCBs and other persistent organic pollutants are lipophilic and are stored in body fat (World Wildlife Fund 1999). They have long half-lives, and are steadily added to the circulation while being slowly metabolized. PCBs are transferred across the placenta to the fetus and nursing is an especially effective way to transfer a large dose from mother to infant. Thus, the adolescents in the sample were exposed *in utero* and through breast milk to lipophilic contaminants consumed before the pregnancy and before the fish warnings.

The focal toxicant, PCBs, is a molecule that can take 209 forms, but only about 100 forms, or congeners, are observed with any regularity in the environment and populations (Kimbrough 1995). The molecule can resemble sex steroids and thyroid hormones to some extent, and the concern is that they will mimic the natural, endogenous hormones at crucial points in the developmental process, creating damage that unfolds with development over the lifespan (Bigsby *et al.* 1999). Some PCBs are thought to have agonistic and others antagonistic effects on endocrine function. They also resemble other compounds that are highly toxic such as dioxin (2,3,7,8-tetrachlorodibenzo-*p*-dioxin) and dibenzofuran (Figure 5.4).

The variety of structures of the PCB congeners, in particular the amount and placement of chlorine substitutions, is thought to be responsible for the range of PCB effects. Thus, for the analyses congeners have been grouped together according to the number of chlorine substitutions in the molecule. Total mono-ortho substitution refers to the sum of all congeners with one substituted carbon at the ortho position of the molecule. The designation 1–4Cl, or 5–9Cl indicates whether the total number of chlorines in the molecule at any position is between 1–4 or 5–9. An advantage to this data analysis procedure is that the distributions of values in congener groups are found to be more normally distributed and appropriate for statistical testing. The values were log transformed when necessary to further normalize the distributions.

The other toxicants measured are: DDE (1,1-dichloro-2,2-bis(chlorophenyl) ethylene) which is a metabolite of DDT (1,1,1-trichloro-2,2-bis(*p*-chlorophenyl) ethane) and thus serves as a measure of past DDT exposure, HCB (hexachloro benzene), and mirex (a pesticide), as well as mercury and lead.

The study sample includes 271 males and females aged 10–17 years of age (Gallo *et al.* 2000). The sample is 52 per cent female and the average age is 13 years. Compared with US youth, the sample average height-for-age percentile is 58 while the weight-for-age percentile average is 88. The average body mass index is 23.4, and total caloric intake averages 3180 Kcal/day. In bivariate analyses body mass index (BMI) is significantly and negatively related to levels of PCBs,

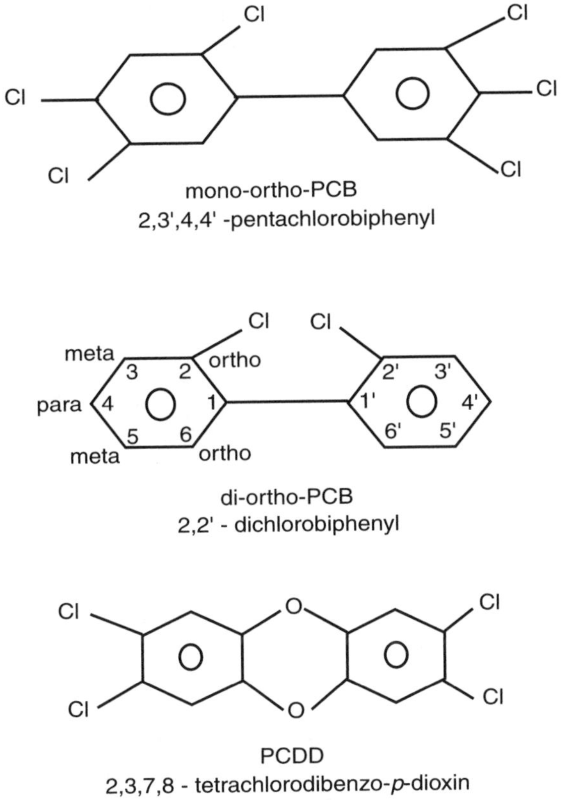

Figure 5.4 Similarity of structures of polyaromatic hydrocarbons: comparison of mono-ortho substituted PCB, di-ortho substituted PCB and 2,3,7,8-tetrachlorodibenzo-*p*-dioxin.

HCB and DDE. Multivariable analysis (Table 5.1) shows the relationship of one group of persistent PCBs to BMI.

When controlling for other influences on BMI such as age, gender and maternal BMI as well macronutrient intakes, BMI is negatively associated with the PCB burden. Other recent reports also have associated BMI with toxicant burdens (Mazhitova *et al.* 1998; Schildkraut *et al.* 1999; Wolff *et al.* 2000), but the finding that a relationship of BMI to toxicants is independent of macronutrients is new. At this stage of the analysis the intercorrelation of other toxicants with PCBs is not taken into account, and in light of the close interrelationship among toxicant burdens, these findings must be considered preliminary and subject to revision.

For analysis of endocrine parameters, a preliminary sample is available consisting of 117 males and females between the ages of 10 and 17 years. (Endocrine analyses for the remaining 154 participants are forthcoming.) The preliminary results show that hormones involved in the thyroid gland activity are altered by the PCB levels present in the sample (Schell *et al.* 2002). Free (unbound) thyroxine and total

Table 5.1 Multivariable analysis of an adolescent's BMI in relation to their persistent PCB levels (ppb). (Adapted from Gallo *et al.* 2002.)

Body Mass Index[a] ($n = 263$; $r^2 = 0.18$)	Standardized coefficients			95% CI	
	β	t	p	Lower bound	Upper bound
Age (yrs)	0.26	4.61	< 0.001	0.02	0.04
Gender (0=male,1=female)	−0.12	−2.14	0.03	−0.10	0.00
Maternal BMI[a]	0.30	5.38	< 0.001	0.20	0.44
Total protein intake (g/day)	−0.23	−1.77	0.08	−0.33	0.02
Total caloric intake (K cal/day)	0.00	0.01	0.99	−0.26	0.27
Total fat intake (g/day)	0.22	1.37	0.17	−0.05	0.30
Σ of 7 persistent PCBs[b]	−0.16	−2.86	0.01	−0.12	−0.02

Notes
[a] BMI = weight(kg)/height(m^2)
[b] Persistent PCBs: IUPAC #74, 99, 118, 138, 153, 180, 187.

thyroxine levels are decreased in relation to the level of total PCBs (Table 5.2) and in relation to the levels of some sub-groupings of specific congeners with similar molecular structures (i.e. total mono-ortho substitution refers to the sum of all congeners with one substituted carbon at the ortho position of the molecule). Thyroid stimulating hormone is elevated. These two findings are internally consistent and reflect a functional hypothalamic–pituitary–thyroid axis. These relationships are also apparent after employing a multivariable model to adjust for other toxicants such as HCB, DDE, mirex and lead as well as age and sex (Table 5.3). These relationships indicate that thyroid hormone economy is affected by exposure to PCBs.

It is likely that these effects are the result of exposure to PCBs long before the thyroid hormones were measured. Consumption of locally caught fish sharply decreased following the advisories in the 1980s by local health agencies warning against local fish consumption. Postnatal consumption by study participants probably was low and the bulk of their PCB burden had come from prenatal and lactational exposure.

Alterations in adolescent growth and thyroid economy are not necessarily a disease, but they may represent the tip of a health pyramid. Studies of cohorts poisoned by persistent organic pollutants such as PCBs and dioxin have shown severe effects on growth, cognitive development and endocrine function (Chen *et al.* 1992; Guo *et al.* 1994a, 1994b, 1995a, 1995b; Yu *et al.* 1997, 2000) indicating that a 'tip of the iceberg' analogy may be appropriate.

These changes in growth and endocrine profiles indicate that industrial pollution may have subtle though important effects on human biology. These effects are hardly confined to urban populations, and indeed are examples of the far-reaching effects of urbanism and its infrastructural needs.

Table 5.2 Relation of an adolescent's PCB levels (ppb; log transformed) and their thyroid hormones: partial correlation coefficients. (Adapted from Schell *et al.* 2002.)

	Correlation coefficient			
	TSH (a)	*Free T4 (b)*	*T4 (b)*	*T3 (b)*
Total PCBs	0.19*	−0.30**	−0.26**	−0.05
PCB congeners with 5–9 chlorines	0.30**	−0.32**	−0.31**	−0.08
Total mono-ortho substitution (5–9Cl)	0.28**	−0.31**	−0.28**	−0.10
Total di-ortho substitution (5–9Cl)	0.31**	−0.34**	−0.32**	−0.06
Total tri/tetra ortho substitution (5–9Cl)	0.19*	−0.24**	−0.23*	−0.06
PCB congeners with 1–4 chlorines	0.01	−0.19*	−0.13	−0.00
Total mono-ortho substitution (1–4 Cl)	0.003	−0.18*	−0.15	−0.02
Total di-ortho substitution (1–4 Cl)	0.02	−0.17	−0.10	−0.02
Total tri/tetra ortho substitution (1–4 Cl)	0.12	−0.19*	−0.17	0.08

Notes
(a) log transformed.   (b) adjusted for age. $* p \leq .05; \; ** p \leq .01$

The role of urbanism in epidemiologic transitions

Urbanism and urbanization have played key roles in creating patterns of disease and death that characterize the stages elucidated by epidemiologic transition theory. The development of sedentary life that marks the onset of urbanism, is related to the first epidemiologic transition (Barrett *et al.* 1998). The impact of sedentism on population growth, dietary variability, contact with domesticated animals (both wanted and unwanted) and on other factors, changed the profile of disease and death. Later, the development of intercontinental trade between relatively small urban populations spread local diseases as pandemics (McNeill 1976). It also created a large, networked population that could initially sustain chronic infectious diseases such as leprosy, and later, when population size and speed of trade allowed, of more rapidly spreading, acute community infections as well (Schell 1988). By 1600, the common European urban pattern involved animal reservoirs of infectious disease in homes and markets; water contaminated by organic waste, graveyards and craft production (textile dying, tanneries, etc.); inadequate sanitation; privately owned water supply; inter-urban trade that funneled pathogens into the cities; and certain customs of dining and hygiene that probably helped to spread disease. Pestilence and famine were common.

The development of the industrial city attracted a large population from rural districts and created a concentrated population whose living and working conditions, food and water supply, sanitation and hygiene were completely inadequate to human well-being and were largely unregulated by government (Flinn 1968; Wohl 1983). Infectious diseases, including tuberculosis, cholera, typhoid, typhus, diphtheria, and pneumonia, were common causes of death. By the early 20th century, many cities were affected by government regulation of food and water supply, housing, working conditions, sanitation, and the health of children in schools (Burnett 1991; Haines 1991; Tarr 1996). In America and perhaps elsewhere, the commodification of hygiene contributed also. The

Table 5.3 Effect of PCB congener groups on thyroid hormones among adolescents of the Akwesasne Mohawk Nation: multivariable analysis.[a] (Adapted from Schell *et al.* 2002.)

Toxicant [b]	Standardized β coefficients $n = 113$			
	TSH[b]	Free T4	T4	T3
Sum of 7 congeners[c]	0.36**	−0.35**	−0.27**	−0.07
PCB congeners with 5–9 Cl	0.23*	−0.28**	−0.24*	−0.07
Mono-ortho substitution (5–9Cl)	0.25*	−0.28**	−0.21*	−0.09
Di-ortho substitution (5–9Cl)	0.25*	−0.30**	−0.25*	−0.06
Di-ortho substitution (1–4Cl)	0.03	−0.19	−0.10	0.05
Tri/tetra ortho substitution (5–9Cl)	0.12	−0.20*	−0.17	−0.04

Notes
* $p < 0.05$ ** $p < 0.01$
[a] Covariates are mirex, hexachlorobenzene, DDE, lead, age and sex.
[b] log transformed.
[c] IUPAC #s: 153, 138+164+163, 118, 180, 187, 170, 183. Structure: 245/245, 236/345+234/ 245+ 2356/34, 245/34, 2345/245, 2356/245, 2345/234, 2346/245.

commercial promotion of many personal hygienic products, i.e. soap and shampoo, created higher standards of personal hygiene for assimilation into mainstream culture and social success (Hoy 1995). Through government and commerce, living conditions became better. Although the causes of the epidemiologic transition that occurred in the US and western Europe at the close of the 19th century are controversial (Barrett *et al.* 1998; Porterfield and Hendry 1998; Rogers and Hackenberg 1987), by the first quarter of the 20th century government regulation was a key factor in making cities livable, that is relatively healthy and not sinks of disease and disability.

The transition at the turn of the 19th century included the growth of government regulation of organic pollution, for example through financing of city-wide sanitation systems (Wohl 1983; Tarr 1996). However, methods for managing organic pollution were not so effective in controlling industrial pollution (Ember 1982) and widespread environmental contamination has resulted.

Further changes in patterns of disease and death may result from human exposure to these contaminants. Past and ongoing industrial pollution has biological effects that bear on evolution and adaptation in both developed and developing countries. Pollution is a slippery slope. The number of tons of pollutants entering the environment each day numbs our sensitivity, and our tolerance for a little more contamination increases accordingly. Certainly few people desire to sacrifice the lifestyle that requires enormous amounts of energy and produces potent amounts of waste. The fastest growing urban areas on Earth are unlikely to sacrifice the traditional benefits of city living for a less contaminated environment. Furthermore, they are without the infrastructure to enforce regulations over pollutants, making increased environmental contamination likely. Conversely, the US, one of the most economically developed countries, will not address the problem having withdrawn as of this writing from a worldwide agreement on a strategy to limit air pollution. If the solution to environmental contamination is government regulation then an increase in the global pollutant burden is likely with consequences for human biology for the foreseeable future.

Conclusion

The study of human urban ecology provides an enormous opportunity to examine bio-cultural interaction because specific challenges to human biology in the urban environment are produced by human activity organized through cultural forms. The dramatic increase in the size of the urban population globally implies that urbanism is a successful adaptive strategy for the species. However, there is considerable evidence that this strategy has spawned additional challenges insofar as urban pollution contributes to human mortality, morbidity and decreased well-being.

References

Alderman, B.W., Baron, A.E. and Savitz, D.A. (1987) 'Maternal exposure to neighborhood carbon monoxide and risk of low infant birth weight', *Public Health Reports*, 102: 410–14.

Barrett, R., Kuzawa, C.W., McDade, T. and Armelagos, G.J. (1998) 'Emerging and re-emerging infectious diseases: the third epidemiologic transition', *Annual Review of Anthropology*, 27: 247–71.

Bentley, G.R. (2000) 'Environmental pollutants and fertility', in G.R. Bentley and C.G.N. Mascie-Taylor (eds) *Infertility in the Modern World: Present and Future Prospects*, pp. 85–152, Cambridge: Cambridge University Press.

Bigsby, R., Chapin, R.E., Daston, G.P., Davis, B.J., Gorski, J., Gray, L.E. Jr., Howdeshell, K.L., Zoeller, R.T. and vom Saal, F.S. (1999) 'Evaluating the effects of endocrine disruptors on endocrine function during development', *Environmental Health Perspectives*, 107: 613–18.

Bobak, M. (2000) 'Outdoor air pollution, low birth weight, and prematurity', *Environmental Health Perspectives*, 108: 173–6.

Bobak, M. and Leon, D.A. (1992) 'Air pollution and infant mortality in the Czech Republic, 1986–88', *Lancet*, 340: 1010–14.

Bobak, M. and Leon, D.A. (1999a) 'The effect of air pollution on infant mortality appears specific for respiratory causes in the postneonatal period', *Epidemiology*, 10: 666–670.

Bobak, M. and Leon, D.A. (1999b) 'Pregnancy outcomes and outdoor air pollution: an ecological study in districts of the Czech Republic 1986–8', *Occupational and Environmental Medicine*, 56: 539–543.

Bogin, B. (1988) 'Rural-to-urban migration', in C.G. Mascie-Taylor and G.W. Lasker (eds) *Biological Aspects of Human Migration*, pp. 90–127, Cambridge: Cambridge University Press.

Borja-Aburto, V.H., Castillejos M., Gold, D.R., Bierzwinski, S. and Loomis, D. (1998) 'Mortality and ambient fine particles in Southwest Mexico City, 1993–1995', *Environmental Health Perspectives*, 106: 849–55.

Burnett, J. (1991) 'Housing and the decline of mortality', in R. Schofield, D. Reher and A. Bideau (eds) *The Decline of Mortality in Europe*, pp. 158–176, Oxford: Clarendon Press.

Carlsen, E., Giwercman, A., Keiding, N. and Skakkebaek, N.E. (1992) 'Evidence for decreasing quality of semen during past 50 years', *British Medical Journal*, 305: 609–13.

Carlsen, E., Giwercman, A., Keiding, N. and Skakkebaek N.E. (1995) 'Declining semen quality and increasing incidence of testicular cancer: is there a common cause?', *Environmental Health Perspectives*, 103: 137–9.

Chen, Y.-C.J., Guo, Y.-L., Hsu, C.-C. and Rogan, W.J. (1992) 'Cognitive development of Yu-Cheng ("oil disease") children prenatally exposed to heat-degraded PCBs', *Journal of the American Medical Association*, 268: 3213–18.

Crisp, T.M., Clegg, E.D., Cooper, R.L., Wood, W.P., Anderson, D.G., Baetcke K.P., Hoffmann, J.L., Morrow, M.S., Rodier, D.J., Schaeffer, J.E., Touart, L.W., Zeeman, M.G. and Patel Y.M. (1998) 'Environmental endocrine disruption: an effects assessment and analysis', *Environmental Health Perspectives*, 106: 11–56.

Danker-Hopfe, H. and Hulanicka, B. (1995) 'Maturation of girls in lead polluted areas', in R. Hauspie, G. Lindgren and F. Faulkner (eds) *Essays on Auxology*, pp. 334–42, Welwyn Garden City: Castlemead Publications.

Dejmek, J., Selevan, S.G., Benes, I., Solansky, I. and Sram R.J. (1999) 'Fetal growth and maternal exposure to particulate matter during pregnancy', *Environmental Health Perspectives*, 107: 475–80.

Department of Economic and Social Affairs (2000) *World Population Monitoring 1999*, New York: United Nations.

Dolk, H., Pattenden, S., Vrijheid, M., Thakrar, B. and Armstrong B.G. (2000) 'Perinatal and infant mortality and low birth weight among residents near cokeworks in Great Britain', *Archives of Environmental Health*, 55: 26–30.

Eberhardt, M., Ingram, D., Makuc, D.M., Pamuk, E.R., Freid, V.M., Harper, S.B., Schoenborn, C.A. and Xia, H. (2001) *The Urban and Rural Health Chartbook*, Hyattsville, MD: National Center for Health Statistics.

Ember, L.R. (1982) 'Love canal: uncertain science, politics, and law', in C. Hohenemser and J.X. Kasperson (eds) *Risk in the Technological Society*, pp. 77–99, Boulder, CO: Westview Press.

Flinn, M.W. (1968) *Public Health Reform in Britain*, New York: MacMillan.

Gallo, M., Ravenscroft J., Schell L. M., DiCaprio A. and Akwesasne Task Force on the Environment (2000) 'Environmental contaminants and growth of Mohawk adolescents at Akwesasne', *Acta Medica Auxologica*, 32: 72.

Gallo, M., Ravenscroft, J., Schell, L.M., DiCaprio, A. and Akwesasne Task Force on the Environment (2002) 'Environmental contaminants and growth of Mohawk adolescents at Akwesasne', in G. Gill, L. Benson and L.M. Schell (eds) *Human Growth From Birth to Maturity*, pp. 279–87, London: Smith Gordon.

Guo, Y.L., Lin, C.J., Yao, W.J., Ryan, J.J. and Hsu, C.C. (1994a) 'Musculoskeletal changes in children prenatally exposed to polychlorinated biphenyls and related compounds (Yu-Cheng children)', *Journal of Toxicology and Environmental Health*, 41: 83–93.

Guo, Y.L., Chen, Y.-C., Yu, M.-L. and Hsu C.-C. (1994b) 'Early development of Yu-Cheng children born seven to twelve years after the Taiwan PCB outbreak', *Chemosphere*, 29: 2395–404.

Guo, Y.L., Lai, T.J., Chen, S.J. and Hsu C.C. (1995a) 'Gender-related decrease in Raven's Progressive Matrices scores in children prenatally exposed to polychlorinated biphenyls and related contaminants', *Bulletin of Environmental Contamination and Toxicology*, 55: 8–13.

Guo, Y.L., Lambert G.H. and Hsu, C.C. (1995b) 'Growth abnormalities in the population exposed *in utero* and early postnatality to polychlorinated biphenils and dibenzofurans', *Environmental Health Perspectives*, 103: 117–22.

Haines, M.R. (1991) 'Conditions of work and the decline of mortality', in R. Schofield, D. Reher and A. Bideau (eds) *The Decline of Mortality in Europe*, pp. 177–95, Oxford: Clarendon Press.

Hoy, S. (1995) *Chasing Dirt. The American Pursuit of Cleanliness*, New York: Oxford University Press.

Jedrychowski, W., Flak, E. and Mroz, E. (1999) 'The adverse effect of low levels of ambient air pollutants on lung function growth in preadolescent children', *Environmental Health Perspectives*, 107: 669–74.

Johansen, B.E (2000) 'Pristine no more', *The Progressive*, 64: 27–9.

Kimbrough, R.D. (1995) 'Polychlorinated biphenyls (PCBs) and human health: an update', *Critical Reviews in Toxicology*, 25: 133–63.

Lee, J-T., Shin, D. and Chung, Y. (1999) 'Air pollution and daily mortality in Seoul and Ulsan, Korea', *Environmental Health Perspectives*, 107: 149–54.

Loomis, D., Castillejos M., Gold, D.R., McDonnell, W.F. and Borja-Aburto, V.H. (1999) 'Air pollution and infant mortality in Mexico City', *Epidemiology*, 10: 118–23.

Mazhitova, Z., Jensen, S., Ritzen, M. and Zetterstrom, R. (1998) 'Chlorinated contaminants, growth and thyroid function in schoolchildren from the Aral Sea region in Kazakhstan', *Acta Paediatrica*, 87: 991–995.

McNeill, W.H. (1976) *Plagues and Peoples*, New York: Anchor Books.

Moolgavkar, S.H. (2000) 'Air pollution and daily mortality in three U.S. counties', *Environmental Health Perspectives*, 108: 777–84.

Penna, M.L.F. and Duchiade, M.P. (1991) 'Air pollution and infant mortality from pneumonia in the Rio de Janeiro Metropolitan Area', *Bulletin of the Pan American Health Organization*, 25: 47–54.

Pereira, L.A.A., Loomis, D., Conceicao, G.M.S., Braga, A.L.F., Arcas, R.M., Kishi, H.S., Singer, J.M., Bohm, G.M. and Saldiva, P.H.N. (1998) 'Association between air pollution and intrauterine mortality in Sao Paulo, Brazil', *Environmental Health Perspectives*, 106: 325–9.

Peters, A., Skorkovsky, J., Kotesovec, F., Brynda, J., Spix, C., Wichmann, H.-E. and Heinrich, J. (2000) 'Associations between mortality and air pollution in Central Europe', *Environmental Health Perspectives*, 108: 283–7.

Porterfield, S.P. and Hendry, L.B. (1998) 'Impact of PCBs on thyroid hormone directed brain development', *Toxicology and Industrial Health*, 14: 103–20.

Ritz, B. and Yu, F. (1999) 'The effect of ambient carbon monoxide on low birth weight among children born in Southern California between 1989 and 1993', *Environmental Health Perspectives*, 107: 17–25.

Rogers, R.G. and Hackenberg, R. (1987) 'Extending epidemiologic transition theory: a new stage', *Social Biology*, 34: 234–243.

Schell, L.M. (1988) 'Cities and human health', in G. Gmelch and W. Zenner (eds) *Urban Life*, pp. 18–35, Prospect Heights, IL: Waveland Press.

Schell, L.M. (1991a) 'Effects of pollutants on human prenatal and postnatal growth: noise, lead, polychlorinated compounds and toxic wastes', *Yearbook of Physical Anthropology*, 34: 157–88.

Schell, L.M. (1991b) 'Pollution and human growth: lead, noise, polychlorobiphenyl compounds and toxic wastes', in C.G. Mascie-Taylor and G.W. Lasker (eds) *Applications of Biological Anthropology to Human Affairs*, pp. 83–116, Cambridge: Cambridge University Press.

Schell, L.M. and Hills, E.A. (2002) 'Polluted environments as extreme environments', in G. Gilli, L.M. Schell and L. Benson (eds) *Human Growth from Birth to Maturity*, pp. 249–61, London: Smith-Gordon.

Schell, L.M. and Tarbell, A.M. (1998) 'A partnership study of PCBs and the health of Mohawk youth: lessons from our past and guidelines for our future', *Environmental Health Perspectives*, 106: 833–40.

Schell, L.M. and Ulijaszek, S.J. (1999) 'Urbanism, urbanisation, health and human biology: an introduction', in L.M. Schell and S.J. Ulijaszek (eds) *Urbanism, Health and Human Biology in Industrialised Countries*, pp. 3–20, Cambridge: Cambridge University Press.

Schell, L.M., Hubicki, L., DiCaprio, A., Gallo, M. and Akwesasne Task Force on the Environment (2002) 'Polychlorinated biphenyls (PCBs) and thyroid function in adolescents of the Mohawk Nation at Akwesasne', in G. Gilli, L.M. Schell and L. Benson (eds) *Human Growth from Birth to Maturity*, pp. 289–96, London: Smith Gordon.

Schildkraut, J.M., Demark-Wahnefried, W., DeVoto, E., Hughes, C., Laseter, J. L. and Newman, B. (1999) 'Environmental contaminants and body fat distribution', *Cancer Epidemiology, Biomarkers and Prevention*, 8: 179–83.

Schlipköter, H.W., Rosicky, B., Dolgner, R. and Peluch, L. (1986) 'Growth and bone maturation in children from two regions of the F.R.G. differing in the degree of air pollution: results of the 1974 and 1984 surveys', *Journal of Hygiene, Epidemiology, Microbiology, and Immunology*, 30: 353–8.

Schmidt, P. and Dolgner, R. (1977) 'Interpretation of some results of studies in schoolchildren living in areas with different levels of air pollution', *Zentralblatt fur Bakteriologie*, 165: 539–47.

Schwartz, J. (1991) 'Particulate air pollution and daily mortality in Detroit', *Environmental Research*, 56: 204–13.

Stinson, S., Bogin, B., Huss-Ashmore, R. and O'Rourke, D. (2000) *Human Biology: An Evolutionary and Biocultural Perspective*, New York: John Wiley & Sons.

Tarr, J.A. (1996) *The Search for the Ultimate Sink: Urban Pollution in Historical Perspective*, Akron: University of Akron Press.

United Nations Environment Program (1991) *Urban air pollution*, Nairobi: United Nations Environment Program.

Waldbott, G.L. (1978) *Health effects of environmental pollutants*, St. Louis: Mosby.

Waskernagel, M. and Rees, W. (1996) *Our Ecological Footprint*, Gabriola Is., BC: New Society Publishers.

Weber, A.F. (1967) *The Growth of Cities in the Nineteenth Century*, Ithaca: Cornell University Press.

Wohl, A.S. (1983) *Endangered Lives. Public Health in Victorian Britain*, Cambridge, MA: Harvard University Press.

Wolff, M.S., Berkowitz, G.S., Brower S., Senie, R., Bleiweiss, I.J., Tartter, P., Pace, B., Roy, N., Wallenstein, S. and Weston, A. (2000) 'Organochlorine exposures and breast cancer risk in New York City women', *Environmental Research*, 84: 151–61.

Woodruff, T.J., Grillo, J. and Schoendorf, K.C. (1997) 'The relationship between selected causes of postneonatal infant mortality and particulate air pollution in the United States', *Environmental Health Perspectives*, 105: 608–12.

Working Group on Public Health and Fossil-Fuel Combustion (1997) 'Short-term improvements in public health from global-climate policies on fossil-fuel combustion: an interim report', *Lancet*, 350: 1341–9.

World Wildlife Fund (1999) 'Persistent organic pollutants: hand-me-down poisons that threaten wildlife and people', Washington, DC: World Wildlife Fund.

Xu, Z., Jing, L., Yu, D. and Xu, X. (2000) 'Air pollution and daily mortality in Shenyang, China', *Archives of Environmental Health*, 55: 115–20.

Yu, M.-L., Guo, Y.L., Hsu, C.-C. and Rogan W.J. (1997) 'Increased mortality from chronic liver disease and cirrhosis 13 years after the Taiwan "Yucheng" ("oil disease") incident', *American Journal of Industrial Medicine*, 31: 172–5.

Yu, M.-L., Guo, Y.-L.L., Hsu, C.-C. and Rogan, W.J. (2000) 'Menstruation and reproduction in women with polychlorinated biphenyl (PCB) poisoning: long-term follow-up interviews of the women from Taiwan Yucheng cohort', *International Journal of Epidemiology*, 29: 672–7.

Zmirou, D., Schwartz, J., Saez, M., Zanobetti, A., Wojtyniak, B., Touloumi, G., Spix, C., Ponce de Leon, A., Le Moullec, Y., Bacharova, L., Schouten, J., Ponka, A. and Katsouyanni, K. (1998) 'Time-series analysis of air pollution and cause-specific mortality', *Epidemiology*, 9: 495–503.

6 Protecting pregnant women from malaria

S. W. Lindsay, R. McGready and
G. E. L. Walraven

Introduction

Malaria during pregnancy is a major threat to the health of mothers and their children, during development *in utero* and early infancy (Brabin, 1991; Menendez, 1995). The severity of this infection depends to a large extent on the degree of exposure to malaria parasites and, consequently, the development of protective immunity in these women (Table 6.1). Since there is a general lowering of immunity during pregnancy, pregnant women are more likely to be infected, have higher parasite densities and experience more clinical episodes of malaria than when not pregnant (Diagne *et al.*, 1997) in high transmission areas. In many cases these infections lead to anaemia and can be an important indirect cause of maternal mortality (Granja *et al.*, 1998). Malaria infections also harm the foetus. Parasites can cause abortions and stillbirths, particularly in areas of low transmission, where immunity is low. More commonly, infection with malaria leads to low birthweight babies, mainly as a result of intra-uterine growth retardation and less often as pre-term birth (Reinhart, 1978; Watkinson *et al.*, 1985; Dolan *et al.*, 1993; Meuris *et al.*, 1993; Garner *et al.*, 1994; Nosten *et al.*, 1994; Steketee *et al.*, 1996a), as well as an increase in early infant mortality (Brabin, 1991; Meuris *et al.*, 1993; Menendez, 1995; Sullivan *et al.*, 1999).

Although the precise nature of the pathophysiology of malaria infections and the resulting immune responses are poorly understood, some general patterns are clear. In areas of low or epidemic transmission, all pregnant women are vulnerable to infection. Where malaria transmission is intense, the harmful effects of malaria during pregnancy are confined largely to women in their first and second pregnancy, before the development of a strong protective immunity (McGregor *et al.*, 1983). One of the reasons for this protection is that a sub-population of parasites that sequester in the placenta initiate the production of antibodies during the first pregnancy that are protective against parasites in subsequent pregnancies (Fried & Duffy, 1996). Consequently, rates of maternal malaria decline accordingly in multiparous women. Treating infected women from malaria endemic areas is difficult since many of these women remain asymptomatic, so do not seek care for malaria (Steketee *et al.*, 1996b), and placental malaria is often not accompanied by a patent peripheral parasitaemia, so infection cannot be reliably screened for

Table 6.1 Strength of clinical responses to malaria during pregnancy

Response	Maternal immune status		
	Non immune	*Low immunity*	*High immunity*
Susceptibility to infection	++++	+++	++
Risk of illness	++++	+++	+
Severe anaemia	?	+++	+++
Severe/cerebral malaria	++++	+++	−
Materno-fetal mortality	++++	+++	+
Low birthweight	?	++	++
Fetal wastage	++++	++++	+
Gravida at risk	All	All	Primigravidae
Placental parasitaemia	?	+	+++

Source: Reproduced from Duffy and Fried (2001) with permission from Taylor & Francis.

antenatally. For both these reasons a strategy is needed to reduce the exposure of pregnant women to malaria parasites.

Increased risk of exposure during pregnancy

Although it is clear that much pathology results from the altered immunological and physiological environment during pregnancy, it was hypothesized that other changes during pregnancy might make pregnant women more attractive to blood-questing mosquitoes, increasing their risk of acquiring an infection. In order to test this, we carried out field studies in The Gambia.

First, the relative attractiveness of 36 pregnant and 36 non-pregnant women to malaria mosquitoes were compared using experimental huts. Each night for three consecutive nights, three pregnant and three non-pregnant women slept alone under a bednet, in six identical huts (Lindsay *et al.*, 2000). The number of mosquitoes collected from each hut the following morning was used to estimate the relative mosquito-attractiveness of each woman. Twice as many *Anopheles gambiae* mosquitoes, the principal African malaria vector, were attracted to pregnant women than their non-pregnant counterparts.

Second, the attractiveness of pregnant and non-pregnant women were compared at close range by collecting blood-fed mosquitoes from under untreated bednets. Although bednets reduce the number of bites individuals receive (Clarke *et al.*, 2001), some mosquitoes manage to find a way under a net, take a blood meal and rest on the netting (Lindsay *et al.*, 1995). Bloodfed mosquitoes were collected from two groups of bednets; in either group a pregnant or non-pregnant woman slept with a child. In this way the child acted as a control, so that direct comparisons could be made between both groups of women. In order to separate the blood meals of mosquitoes that had fed on such closely related individuals, an improved technique for DNA-fingerprinting the bloodmeal of mosquitoes collected from the field was developed (Ansell *et al.*, 2000). The genotype of each mosquito's meal was then compared with that of the mother or child sleeping under the net. The results showed that pregnant women were two to four times more likely to be bitten by *An. gambiae* mosquitoes than non-pregnant women under a bednet. Moreover, pregnant women also received proportionately more bites under a bednet than non-pregnant women (70 per cent *vs* 52 per cent). Both studies illustrated that pregnant women were more exposed to malaria parasites than other women, a factor that contributes to the greater vulnerability of women to malaria during pregnancy.

In order to determine the reason for the increased attractiveness, 27 pregnant and 27 non-pregnant Gambian women were examined for a variety of factors thought to be important for host location by mosquitoes (K. Hamilton, J. Ansell and S. Lindsay, unpublished data). The findings suggest that pregnant women attracted more mosquitoes due to a combination of factors; a greater volume of exhaled breath (21 per cent), an increase in body surface temperature (0.7 °C) (Ansell *et al.*, 2000), reduced skin sensitivity, and, perhaps, the greater release of volatiles from the skin. Pregnant women will be exposed more, not only because of these direct physiological differences, but also because they leave their bednets twice as often during the night than non-pregnant women. Presumably this is a result of the increased frequency of urination during pregnancy (Brunner *et al.*, 1994). This activity results in women exposing themselves more often to nocturnal vectors outdoors and the frequent opening and closing of the sides of the bednet make it easier for mosquitoes to enter the net.

Treatment problems

Pregnant women can be protected against malaria by chemoprophylaxis and intermittent treatment (Phillips-Howard, 1999). However, the efficacy and sustainability of this practice is limited by the unrelenting spread of parasite strains resistant to antimalarials, lack of compliance with regular chemoprophylaxis, as well as logistical, social (Robb, 1999) and political constraints. Previous studies conducted on pregnant women have shown that by preventing or treating malaria in pregnancy, anaemia can be reduced and birthweight improved (Greenwood *et al.*, 1989). Weekly prophylaxis with chloroquine has been the mainstay of prevention in many countries but has not been adhered to well (Heymann *et al.*, 1990), probably because of its bitter taste and women forget to take it weekly. In Tanzania failure to use chloroquine prophylaxis amongst pregnant women was due to the perceived lack of protection against malaria and fear of chloroquine-induced pruritus (Mnyika *et al.*, 1995). Because of the low compliance with chloroquine prophylaxis and advancing strains of chloroquine-resistant parasites, studies conducted in Kenya and Malawi were undertaken to address the effectiveness of intermittent treatment with antimalarials. These studies demonstrated that intermittent treatment with sulfadoxine-pyremethamine (SP) given in the second and third trimesters can reduce placental parasitaemia (Schultz *et al.*, 1994), severe maternal anaemia (Shulman *et al.*, 2001) and improve birthweight (Parise *et al.*, 1998). This regime appears safe and can be delivered effectively through antenatal clinics. It is now policy in Malawi, Kenya and Tanzania, although resistance against SP is increasing rapidly in these countries.

A good illustration of what the future may hold is found on the Thai-Burmese border, an area of low transmission with a high level of multiple-drug resistant malaria strains. Here there are no safe and effective drugs available for chemo-prophylaxis during pregnancy (McGready & Nosten, 1999). Around 33 per cent of uncomplicated *Plasmodium falciparum* cases treated with quinine fail (McGready *et al.*, 2000) and resistance to SP and mefloquine have made them ineffective on their own. Moreover other antimalarials like halofantrine, tretracycline, doxycycline and primaquine are contraindicated in pregnancy (Phillips-Howard and Wood, 1996), leaving few alternatives. In this setting, the rapid detection of infections by weekly screening and prompt treatment eliminated maternal mortality but failed to completely prevent low birthweight babies and maternal anaemia (Nosten *et al.*, 1991). At present, antimalarial combinations with artemisinins are extremely effective at reducing clinical malaria in this area, but as resistance has developed to all antimalarials to date (White, 1992), given time, resistance to artemisinins will occur. Using drug combinations is likely to delay the process, but the remaining stock of alternative treatments is perilously low. The world is still waiting for an effective vaccine to be produced, but its use in pregnancy will be even longer in appearing since developing a malaria vaccine for pregnant women raises important questions about the impact of immunity for the mother and foetus. Although malaria during pregnancy is harmful, it may be beneficial to the foetus by increasing resistance to malaria infection in infancy (Snow *et al.*, 1997; Menendez, 1999). For these reasons, there is need to consider how to improve protection against malaria during pregnancy.

Indoor spraying

It has been repeatedly demonstrated that indoor-spraying with insecticides can protect communities from malaria in many parts of the tropics (Kouznetsov, 1977). However, their long-term use is limited where home-owners refuse entry of spraymen into their homes, where a significant proportion of the vector population bites outdoors and where the rise of insecticide-resistant strains of mosquito can render this strategy ineffective. Although it has been shown that indoor spraying can reduce morbidity and mortality from malaria in children, it is perhaps surprising that there have been no studies carried out to evaluate indoor spraying for protecting pregnant women from malaria. There is also little data about the safety of many insecticides during pregnancy.

Bednets and protection

As a result of the success of large-scale intervention trials of insecticide-impregnated bednets (IBNs) in Africa (Lengeler, 1998), this strategy now forms one of the central planks for the control of malaria in the tropics. Since it was demonstrated that IBNs were extremely effective at reducing child mortality, it was logical that this intervention should be assessed for protecting pregnant women. However, the studies reported in the literature are not as effective as one might hope. To date it appears that the greatest protection is obtained only in areas where transmission is low and that the nets fail to work in sites where exposure is intense (Table 6.2). This may reflect a higher prevalence of low-grade chronic infections in women living in areas of intense transmission or a failure to adequately treat infected women. When these women become pregnant the density of parasitaemia rises, resulting in the serious pathology associated with malaria in pregnancy. If this theory is correct, treatment with effective antimalarial drugs in early pregnancy to clear established parasite infections and the use of IBNs to prevent new infections should prove to be an extremely successful strategy in the short term.

Mosquito repellents

Protecting pregnant women against malaria on the Thai-Burmese border presents many problems. Not only are there few treatments available for the treatment of malaria, but the efficacy of IBNs is also hindered by the biting activity of local mosquito vectors. Here a substantial amount of biting by the main vectors, *An. maculatus* (66 per cent) and *An. minimus* (22 per cent), occurs outdoors, early in the evening, when people are outside their nets (Lindsay *et al.*, 1998). A series of studies was carried out to investigate whether mosquito repellents could protect pregnant women against malaria in this area. A pilot study showed that repellents used in the early evening gave long-lasting protection against nuisance mosquitoes (Lindsay *et al.*, 1998). A mixture of DET (*N,N*-diethyl-meta-toluamide) and, the local cosmetic, thanakha (*Limonia acidissima*), applied as a paste to the skin gave good protection for around eight hours.

To test whether this repellent mixture protected pregnant women from malaria a double-blind randomised therapeutic trial (McGready *et al.*, 2001) was undertaken. A total of 897 pregnant women were recruited to this study, of whom 449 had DET and thanaka daily and 448 had thanaka alone. Compliance was extremely high, with 93 per cent of women attending the weekly ante-natal clinics, 91 per cent reported using the treatment daily and 85 per cent were using the treatment when visited unannounced at home by one of the survey team. There was a 28 per cent reduction in the incidence of falciparum malaria, the most dangerous form of malaria, and a 20 per cent reduction in vivax malaria in women using the repellent mixture, although these differences were not statistically significant. The inability to demonstrate a statistically significant effect is in part related to the extremely low level of malaria in the study area as a result of the large-scale use of mefloquine together with artesunate (Nosten *et al.*, 2000) and degradation of mosquito-breeding sites. Moreover, thanaka alone is also a weak mosquito repellent (Lindsay *et al.*, 1998), so the trial was not a simple comparison between a repellent and a non-repellent.

One concern about using DET in this trial was related to its safety to the mother and developing child. Although DET is the most widely used repellent in the world (Fradin, 1998) and is considered safe (WHO, 1991; Goodyer & Behrens, 1998) formulations of the repellent carry labels not recommending their use during

Table 6.2 Impact of insecticide-treated bednets on malaria in pregnancy

Country	Transmission intensity	Outcome measurement	Source
Thailand	Low	> 48% reduction in anaemia	Dolan *et al.*, 1993
The Gambia	Low to intense seasonal	39% reduction in parasite prevalence, 67% fewer premature babies and heavier babies	D'Allessandro *et al.*, 1996
Kenya	Moderate	25% reduction in severe malaria (but not significant)	Shulman *et al.*, 1998
Ghana	Moderate	No impact on maternal anaemia	Browne *et al.*, 2001

pregnancy. Surprisingly there have been no studies that have assessed the safety of DET in human pregnancy or infancy, although there was one reported case of an adverse reaction after DET application in a pregnant woman (Schaefer & Peters, 1992). Thus it was clearly important to determine whether the repellent mixture was safe or not. Women in the trial applied substantial quantities of DET to their bodies (median dose 214 g/pregnancy) and it was found that amongst heavy users DET crossed the placenta in 8 per cent of pregnancies (McGready *et al.*, 2002). Despite this high and prolonged exposure there were no adverse neurological, gastrointestinal or dermatological effects seen in the group of women given DET and thanaka compared with those with thanaka alone. Nor were there any adverse affects on survival, growth and development seen in infancy. Clearly DET is safe for use in the second and third trimesters of pregnancy. Taken together, these results are encouraging and suggest that repellents may have a role to play for protecting pregnant women in areas of low transmission in combination with other methods of protection.

Other solutions

In many parts of the tropics, pregnant women remain unaware of the dangers of malaria during pregnancy. It is important that all women receive health education about this life-threatening infection and that they understand what they can do to protect themselves. This should not be left to medical staff at ante-natal clinics alone, but should start at school and at home. Part of the reason for the successful completion and high compliance in the repellency trial in Thailand resulted from these women being acutely aware of the dangers of malaria in pregnancy, high accessibility to health services and the high standard of medical provision in this setting.

Since there are so few tools and limited interest and resources for protecting pregnant women against malaria, the search for other protective mechanisms that might contribute significantly to a reduction in transmission needs to be innovative.

Here are just a few examples of some of the possibilities that could be considered. Few studies have assessed the efficacy of local repellents against mosquito bites and yet the enormously diverse nature of secondary plant products produced by flora, that have evolved as a protection against plant-feeding insects, is a rich source of potential products. If skin bacteria are producing volatiles that attract mosquitoes out for blood, then antibacterial skin treatments may restrict the production of these volatiles, making pregnant women less vulnerable to attack from mosquitoes. Recent studies in The Gambia have demonstrated that untreated bednets are protective against malaria only if the nets are in good condition (Clarke *et al.*, 2001). However, even nets with five small holes big enough to poke a finger through result in an increased risk of malaria infection in young children. It is tempting to think that the simple and cheap intervention of a needle and thread would be a useful method of personal protection against malaria.

Conclusion

Pregnant women are extremely vulnerable to malaria. The increased dangers posed by this disease during pregnancy results from changes in the woman's immunity and physiology, including an increased attractiveness to malaria-carrying mosquitoes. The ability to ameliorate the harmful effects of malaria during pregnancy is compromised by the spread of drug-resistant strains of parasite, having few antimalarials available for treating pregnant women and the lack of an effective vaccine. Protecting pregnant women against mosquitoes will never be the complete answer to this problem, yet anti-vector methods are a useful and essential part of many malaria control programmes. It seems likely that in most malaria-endemic areas, where antifolate drug resistance is not a problem, a combination of antimalarial treatment with one or two doses of SP, perhaps combined with artemisinins, during pregnancy, to eliminate chronic infections, and the use of IBNs, to prevent new infections, would be an effective way of protecting this vulnerable group of women. An attractive feature of this approach is that the strategy can be initiated through an established, and often well-utilised health system of ante-natal clinics. Whilst there is an urgent need to evaluate this combined intervention now, the possibilities of developing alternative methods of protecting this high-risk group of women should also be explored. Most importantly, more government health services across the tropics need to initiate malaria-control strategies specifically targeted at pregnant women using the best available tools.

References

Ansell, J., Hu, J.-T., Gilbert, S. C., Hamilton, K. A., Hill, A. V. S. and Lindsay, S. W. (2000) 'Improved method for distinguishing the human source of mosquito bloodmeals between close family members', *Transactions of the Royal Society of Tropical Medicine and Hygiene*, 94: 572–774.

Brabin, B. (1991) *The Risks and Severity of Malaria in Pregnancy*, World Health Organization, Geneva.

Browne, E., Maude, G. and Binka, F. (2001) 'The impact of insecticide-treated bednets on malaria and anaemia in pregnancy in Kassena-Nankana district, Ghana: a randomized controlled trial', *Tropical Medicine and International Health*, 6: 667–76.

Brunner, D.P., Munch, M., Biedermann, K., Huch, R., Huch, A. and Borbely, A.A. (1994) 'Changes in sleep and sleep electroencephalogram during pregnancy' *Sleep*, 17: 576–82.

Clarke, S.E., Bøgh, C., Brown, R.C., Pinder, M., Walraven, G.E.L. and Lindsay, S.W. (2001) 'Do untreated bednets protect against malaria?' *Transactions of the Royal Society of Tropical Medicine and Hygiene*, 95: 457–62.

D'Allessandro, U., Langerock, P., Bennet, S., Francis, N., Cham, K. and Greenwood, B.M. (1996) 'The impact of a national impregnated bed net programme on the outcome of pregnancy in primigravidae in The Gambia', *Transactions of the Royal Society of Tropical Medicine and Hygiene*, 90: 487–92.

Diagne, N., Rogier, C., Cisse, B. and Trape, J.F. (1997) 'Incidence of clinical malaria in pregnant women exposed to intense perennial transmission', *Transactions of the Royal Society of Tropical Medicine and Hygiene*, 91: 166–70.

Dolan, G., Kuile, F.O.T., Jacoutot, V., White, N.J., Luxemburger, C., Malankirii, L., Chonsuphajaisiddhi, T. and Nosten, F. (1993) 'Bed nets for the prevention of malaria and anaemia in pregnancy', *Transactions of the Royal Society of Tropical Medicine and Hygiene*, 87: 620–6.

Duffy, P. and Fried, M. (2001) *Malaria in Pregnancy*, London: Taylor & Francis.

Fradin, M. (1998) 'Mosquitoes and mosquito repellent', *Annals of Internal Medicine*, 128: 931–40.

Fried, M. and Duffy, P.E. (1996) 'Adherence of *Plasmodium falciparum* to chondroitin sulfate A in the human placenta', *Science*, 272: 1502–4.

Garner, P., Dubowitz, L., Baea, M., Lai, D., Dubowitz, M. and Heywood, P. (1994) 'Birthweight and gestation of village deliveries in Papua New Guinea', *Journal of Tropical Paediatrics*, 40: 37–40.

Goodyer, L. and Behrens, R. (1998) 'Short report: the safety and toxicity of insect repellents', *American Journal of Tropical Medicine and Hygiene*, 59: 323–4.

Granja, A.C., Machungo, F., Gomes, A., Bergstrom, S. and Brabin, B. (1998) 'Malaria-related maternal mortality in urban Mozambique', *Annals of Tropical Medicine and Parasitology*, 92: 257–63.

Greenwood, B.M., Greenwood, A.M., Snow, R.W., Byass, P., Bennet, S. and Hatib-N'jie, A.B. (1989) 'The effects of malaria chemoprophylaxis given by traditional birth attendants on the course and outcome of pregnancy', *Transactions of the Royal Society of Tropical Medicine and Hygiene*, 83: 589–594.

Heymann, D.L., Steketee, R.W., Wirama, J.J., Mcfarland, D.A., Khoromana, C.O. and Campbell, C.C. (1990) 'Antenatal chloroquine chemoprophylaxis in Malawi: chloroquine resistance, compliance, protective efficacy and cost', *Transactions of the Royal Society of Tropical Medicine and Hygiene*, 84: 496–8.

Kouznetsov, R. L. (1977) 'Malaria control by application of indoor spraying of residual insecticides in tropical Africa and its impact on community health', *Tropical Doctor*, 7: 81–91.

Lengeler, C. (1998) 'Insecticide treated bednets and curtains for malaria control (Cochrane Review)', *Cochrane Library*, Issue 3: 1–70.

Lindsay, S.W., Armstrong Schellenberg, J.R.M., Zeiler, H.A., Daly, R.J., Salum, F.M. and Wilkins, H.A. (1995) 'Exposure of Gambian children to *Anopheles gambiae* malaria vectors in an irrigated rice production area', *Medical and Veterinary Entomology*, 9: 50–8.

Lindsay, S.W., Ewald, J.A., Samung, Y., Apiwathnasorn, C. and Nosten, F. (1998) 'Thanaka (*Limonia acidissima*) and deet (di-methyl benzamide) mixture as a mosquito repellent for use by Karen women', *Medical and Veterinary Entomology*, 12: 295–301.

Lindsay, S.W., Ansell, J., Selman, C., Cox, V., Hamilton, K. and Walraven, G.E.L. (2000) 'Effect of pregnancy on exposure to malaria mosquitoes', *The Lancet*, 355: 1972.

McGready, R. and Nosten, F. (1999) 'The Thai-Burmese border: drug studies of Plasmodium falciparum in pregnancy', *Annals of Tropical Medicine and Parasitology*, 93: S19–S23.

McGready, R., Brockman, A., Cho, T., Cho, D., Van Vugt, M., Luxemburger, C., Chongsuphajaisiddhi, T., White, N.J. and Nosten, F. (2000) 'Randomized comparison of mefloquine-artesunate versus quinine in the treatment of multidrug-resistant falciparum malaria in pregnancy', *Transactions of the Royal Society of Tropical Medicine and Hygiene*, 94: 689–93.

McGready, R., Simpson, J.A., Htway, M., White, N.J., Nosten, F. and Lindsay, S.W. (2001) 'A double-blind randomized therapeutic trial of insect repellents for the prevention of malaria in pregnancy', *Transactions of the Royal Society of Tropical Medicine and Hygiene*, 95: 137–8.

McGready, R., Hamilton, K.A., Simpson, J.A., Cho, T., Luxemburger, C., Edwards, R.A., Looareesun, S., White, N.J., Nosten, F. and Lindsay, S.W. (2002) 'Safety of the insect repellent *N,N*-diethyl-*m*-toluamide (DET) in pregnancy', *American Journal of Tropical Medicine and Hygiene*, 65: 285–9.

McGregor, I.A., Wolson, M.E. and Billewicz, W.Z. (1983) 'Malaria infection of the placenta in The Gambia, West Africa; its incidence and relationship to stillbirth, birthweight and placental weight', *Transactions of the Royal Society of Tropical Medicine and Hygiene*, 77: 232–44.

Menendez, C. (1995) 'Malaria during pregnancy: a priority area of malaria research and control', *Parasitology Today*, 11: 178–83.

Menendez, C. (1999) 'Priority areas for current research on malaria during pregnancy', *Annals of Tropical Medicine and Parasitology*, 93: S71–S74.

Meuris, S., Piko, B.B., Eerens, P., Vanbellinghen, A.M., Dramaix, M. and Hennart, P. (1993) 'Gestational malaria: assessment of its consequences on foetal growth', *American Journal of Tropical Medicine and Hygiene*, 48: 603–9.

Mnyika, K.S., Kabalimu, T.K. and Lugoe, W.L. (1995) 'Perception and utilisation of malaria prophylaxis among pregnant women in Dar es Salaam, Tanzania', *East African Medical Journal*, 72: 431–5.

Nosten, F., Kuile, F.T., Maelankirri, L., Decludt, B. and White, N.J. (1991) 'Malaria during pregnancy in an area of unstable endemicity'. *Transactions of the Royal Society of Tropical Medicine and Hygiene*, 85: 424–9.

Nosten, F., Kuile, F.T., Maelankiri, L., Chongsuphajaisiddhi, T., Nopdonrattakoon, L., Tangkichot, S., Boudreau, E., Bunnag, D. and White, N.J. (1994) 'Mefloquine prophylaxis prevents malaria during pregnancy: a double blind, placebo controlled study', *Journal of Infectious Diseases*, 169: 595–603.

Nosten, F., Van Vugt, M., Price, R., Luxemburger, C., Thway, K., Brockman, A., McGready, R., Ter Kuile, F., Looareesuwan, S. and White, N.J. (2000) 'Effects of artesunate-mefloquine combination on incidence of *Plasmodium falciparum* malaria and mefloquine resistance in western Thailand: a prospective study', *The Lancet*, 356: 297–302.

Parise, M.E., Ayisi, J.G., Nahlen, B.L., Schultz, L.J., Roberts, J.M., Misore, A., Muga, R., Oloo, A.J. and Steketee, R.W. (1998) 'Efficacy of sulfadoxlne-pyrimethamine for prevention of placental malaria in an area of Kenya with a high prevalence of malaria and human immunodeficiency virus infection', *American Journal of Tropical Medicine and Hygiene*, 59: 813–22.

Phillips-Howard, P.A. (1999) 'Epidemiological and control issues related to malaria in pregnancy', *Annals of Tropical Medicine and Parasitology*, 93: S11–S17.

Phillips-Howard, P.A. and Wood, D. (1996) 'The safety of antimalarial drugs in pregnancy', *Drug Safety*, 14: 131–45.

Reinhart, M.C. (1978) 'Malaria at delivery in Abidjan – its influence on placenta and newborns', *Helvetica Paediatrica*, 33: S43–S63.

Robb, A. (1999) 'Malaria and pregnancy: the implementation of interventions', *Annals of Tropical Medicine and Parasitology*, 93: S67–S70.

Schaefer, C. and Peters, P. (1992) 'Intrauterine diethyltoluamide exposure and fetal outcome', *Reproductive Toxicology*, 16: 175–6.

Schultz, L., Stekette, R.W., Macheso, A., Kazembe, P., Chitsulo, L. and Wirima, J.J. (1994) 'The efficacy of antimalarial regimens containing sulfadoxine-pyrimethamine and/or chloroquine in preventing peripheral and placental *Plasmodium falciparum* infection among pregnant women in Malawi', *American Journal of Tropical Medicine and Hygiene*, 51: 515–22.

Shulman, C.E., Dorman, E.K., Talisuna, A.O., Lowe, B.S., Nevill, C., Snow, R.W., Jilo, H., Peshu, N., Bulmer, J.N., Graham, S. and Marsh, K. (1998) 'A community randomised controlled trial of insecticide-treated bednets for the prevention of malarial and anaemia among primigravid women on the Kenyan coast', *Tropical Medicine and International Health*, 3: 197–204.

Shulman, C.E., Marshall, T., Dorman, E.K., Bulmer, J.N., Cutts, F., Peshu, N. and Marsh, K. (2001) 'Malaria in pregnancy: adverse effects on haemoglobin levels and birthweight in primigravidae and multigravidae', *Tropical Medicine and International Health*, 6: 770–8.

Snow, R.W., Omumbo, J.A., Lowe, B., Molyneux, C.S., Obiero, J.O., Palmer, A., Weber, M.W., Pinder, M., Nahlen, B., Obonyo, C., Newbold, C., Gupta, S. and Marsh, K. (1997) 'Relation between severe malaria morbidity in children and level of *Plasmodium falciparum* transmission in Africa', *The Lancet*, 349: 1650–4.

Steketee, R.W., Wirima, J.J., Hightower, A.W., Slutsker, L., Heymann, D.L. and Breman, J.G. (1996a) 'The effect of malaria and malaria prevention in pregnancy on offspring birthweight, prematurity, and intrauterine growth retardation in rural Malawi', *American Journal of Tropical Medicine and Hygiene*, 55: 33–41.

Steketee, R.W., Wirima, J.J., Slutsker, L., Breman, J.G. and Heymann, D.L. (1996b) 'Comparability of treatment groups and risk factors for parasitemia at the first antenatal clinic visit in a study of malaria treatment and prevention in pregnancy in rural Malawi', *American Journal of Tropical Medicine and Hygiene*, 55: S17–S23.

Sullivan, A.D., Nyirenda, T., Cullinan, T., Taylor, T., Harlow, S.D., James, S.A. and Meshnick, S.R. (1999) 'Malaria infection during pregnancy: intrauterine growth retardation and preterm delivery in Malawi', *Journal of Infectious Diseases*, 179: 1580–3.

Watkinson, M., Rushton, D.I. and Lunn, P.G. (1985) 'Placental malaria and foetoplacental function: low plasma oestradiols associated with malarial pigmentation of the placenta', *Transactions of the Royal Society of Tropical Medicine and Hygiene*, 79: 448–50.

White, N.J. (1992) 'Antimalarial drug resistance: the pace quickens', *Journal of Antimicrobial Chemotherapy*, 30: 571–85.

WHO (1991) *Safe Use of Pesticides*, World Health Organisation, Geneva.

7 Interdisciplinary research on *Schistosoma japonicum*

Stephen T. McGarvey, Gemiliano D. Aligui,
Jonathan D. Kurtis, Arve Lee Willingham III,
Hélène Carabin, Remigio Olveda

Introduction

Schistosomiasis is one of the six diseases of most concern to the World Health Organization and among human parasitic diseases ranks only behind malaria in terms of socioeconomic and public health importance in tropical and subtropical areas. There are an estimated 600 million individuals residing in schistosomiasis endemic regions with approximately 200 million infected at any time (WHO 1993). *Schistosoma japonicum* is endemic to the Philippines, the People's Republic of China, and Indonesia. It is unique among the major schistosomes infecting humans in that its intermediate host snail is amphibious and it is the only schistosome for which zoonotic transmission is considered important with domesticated animals serving as important reservoir hosts of the parasite (Jordan *et al.* 1993). About 69–75 million individuals are at risk of infection from *S. japonicum*, including approximately 1.5 million current infections (Chitsulo 1998).

The purpose of this chapter is to describe and interpret the results from several community-based studies on *S. japonicum* in the Philippines and China, and to present the rationales for two new interdisciplinary research projects in the Philippines. This is not an exhaustive review of all studies on *S. japonicum* done by the many research groups and individuals in the endemic nations. The emphasis is on the insights obtained from the studies conducted in the 1980s and 1990s by our research group and how they led to new research with an explicit human ecological and interdisciplinary set of perspectives.

The ecology and transmission of schistosomiasis in the Philippines and China results from the interaction of the parasite *S. japonicum* with its complex life cycle, human and other mammalian definitive hosts, regular exposure of hosts to the cercaria, the infective form of the parasite in fresh water, tropical or semi-tropical rainfall and temperature patterns, all within the context of rural poverty and the lack of sanitation systems (Jordan *et al.* 1993). Faeces of infected humans and other mammals contain *S. japonicum* eggs, which hatch in water into miracidia and then infect the intermediate host snail species. The cercerial form emerges from infected snails, penetrates the skin of potential hosts and after complex developmental changes and migration within hosts over several weeks, adult worms take up residence in the mesenteric veins. Male and female worms find one another and the female produces eggs, although *S. japonicum* egg laying is thought to be irregular and clustered. This day-to-day variation in the presence and amount of *S. japonicum* eggs in human faeces, especially in those with light intensity infections, presents challenges to accuracy of detecting infections (Aligui 1997).

Rice farming with irrigation canals and water-filled paddies produces ideal habitats for snails. Humans in these poor farming communities do not have sanitary systems for faecal disposal and contaminate the water with schistosome eggs, which infect the snails. *S. japonicum* -infected mammals also contaminate the environment with their faeces.

S. japonicum transmission in the Philippines and China has been known to include other mammals as definitive hosts but until the last few years the specific contribution of non-human definitive hosts to transmission has not explicitly been studied (McGarvey *et al.* 1999). These non-human hosts include the domesticated animals, carabao (water buffalo), cattle, dogs, pigs, cats, sheep and goats, and wild mammals, especially field rats. As will be described later, this greater appreciation for the potential role of all definitive host species in *S. japonicum* transmission helped interpret findings from earlier studies of infection and disease control, as well as providing motivation for new studies.

In 1990, the Philippine National Schistosomiasis Control Programme under the Philippine Health Development Plan, received substantial World Bank loans. Infection control teams were doubled and almost 100 per cent of endemic villages received regular case-finding and treatment. These funds lasted until 1995 when 50 per cent of the manpower was lost. Thus, population coverage reverted to its pre-intervention level. In 1995, mass chemotherapy on an annual or biennial basis was adopted if community prevalence was greater than 10 per cent. The inherent difficulties in the sustainability of annual case-finding and treatment and perhaps mass treatment strategies (in combination with the opportunities for the development of praziquantel resistance with its continued use) are serious issues. There are no detailed studies evaluating the effectiveness of those infection control activities in endemic areas of the Philippines and no accurate estimates of the community or age-specific prevalence or intensity of infection, or related morbidity.

Over the last two decades, schistosomiasis japonica has been eradicated in the coastal plains region of the People's Republic of China by way of comprehensive programmes involving snail control, large-scale chemotherapy of the human population, and public health education campaigns (Sleigh *et al.* 1998a, 1998b, 1998c; Yuan *et al.* 2000). Despite this success, transmission continues in the lake and marsh regions in central China, the middle and lower Changjian River (Yangtze) and the mountainous areas in Sichuan and Yunnan Provinces. The World Bank loans to China provided also for further research on regional-specific methods to reduce transmission, most involving environmental interventions to reduce transmission (Yuan *et al.* 2000). There remains a significant potential for *S. japonicum* transmission as demonstrated by research on the impact of the severe flooding in 1998 (Bergquist *et al.* 2000).

Our perspective on the current epidemiologic status of *S. japonicum* is that amidst the overall decrease in prevalence, infection and associated disease are becoming more localized within ecological, geographic and socio-economic strata in both the Philippines and China. The following section summarizes the community studies performed by our group in the 1980s and early 1990s, and concentrates on the work in the Philippines where our new research is starting.

Community-level infection control and schistosomiasis-associated morbidity

Impact of annual case-finding and treatment

In the 1980s and throughout the early 1990s, a series of long-term multi-disciplinary studies in the Philippines and China showed substantial schistosomiasis japonica-related morbidity (McGarvey *et al.* 1992, 1995, 1996; Olveda *et al.* 1996; Olds *et al.* 1996; Wiest *et al.* 1992, 1993, 1994). The Philippine studies took place in several villages in northeastern Leyte in a schistosomiasis endemic region with very high rainfall and poverty where earlier schistosomiasis research had taken place (Pesigan *et al.* 1958; Olveda *et al.* 1983; Blas *et al.* 1988). The Chinese study took place in the lake and marsh region where schistosomiasis remains at endemic levels on an island community in Poyang Lake, Jiangxi Province. The basic design of the community interventions was case detection and chemotherapy with praziquantel (the standard anti-schistosomal drug with oral doses based on body weight). Case detection was based on one stool sample and duplicate slide readings using the Kato–Katz techniques (Peters and Kazura 1987). Each year almost all infected individuals were treated. Some did not adhere to the call for treatment and some were lost to follow-up during the three to five days in each village.

In the Philippines this annual case-finding and treatment occurred from 1981 to 1988 and detailed findings for the three villages were reported (Olveda *et al.* 1996). The general trend was for a sharp decrease in prevalence in the first two years, especially in children, adolescents and young adults. This was accompanied by marked reductions in intensity of infection as well. However, despite these reductions in prevalence and intensity of infection after several years of annual case-finding and treatment, the total community prevalence appeared to stabilize at plateau levels of 15–30 per cent, depending on ecological conditions and social and geographic isolation (Olveda *et al.* 1996). This is clearly disappointing but points to the complex ecological and socio-economic context in which the inter-action of this human population and this parasitic infection occurs.

In the Chinese island community, with an assumed high transmission potential due to the daily contact with lake water, a similar trend was found, although the annual case-finding and treatment lasted only from 1989 to 1992 (Wiest *et al.* 1992, 1993, 1994). The age-specific prevalence of children and adolescents was notably higher in this community and it was reduced quite quickly after only one year and continued to decline until the final year. Adult prevalence also decreased, as expected. The intensity of infection also decreased within the first year. However, the prevalence after three rounds of annual case-finding and treatment remained moderate despite the reduction in intensity and associated morbidity (Wiest *et al.* 1994).

Sensitivity of single stool examination

Annual case-finding and treatment based on a single stool sample leads to false negative diagnoses and underestimations of schistosomiasis japonica prevalence. It is assumed that this failure occurs mostly among low intensity infections due to both technical error in recognizing one or few eggs in a Kato–Katz thick smear and to the sporadic egg laying pattern of *S. japonicum* females. This diagnostic imprecision allows individuals with a false negative diagnosis to contaminate the environment, thus continuing transmission and making infection control more difficult. This may have contributed to the moderately high plateau prevalence of infection that was found in the Philippine and Chinese studies described above. In a yet unpublished study, a sample of 224 residents of a village in Leyte provided five stool samples and two slides per sample were prepared by the Kato–Katz thick smear technique (Aligui 1997). A Bayesian statistical approach was employed by taking into consideration *a priori* knowledge and beliefs about the probability distribution of the sensitivity and specificity of the single-stool Kato–Katz exam. Whereas prevalence was estimated at 46 per cent with one stool sample, three and five stool samples yielded prevalence estimates of 64 per cent and 75 per cent, respectively. The Bayesian model showed that using a single stool resulted in the failure to detect 47 per cent of the truly lightly infected, 1–100 eggs per gram (epg) stool, and about 23 per cent of those with true moderate intensity infections, 101–400 epg. Overall, one stool sample underestimated true prevalence by as much as 25–35 per cent. For example, if one stool sample led to a community prevalence estimate of 28 per cent, the true prevalence was approximately 44 per cent. In addition, the 95 per cent confidence intervals for the prevalence estimates decreased with increasing number of stools. In the context of increasing costs of multiple stool collection, preparation and examination, three stool samples provided the optimal prevalence estimates in this endemic community.

Age-dependent reinfection

Eight-year longitudinal data from the Philippines provided suggestive evidence for the presence of age-dependent acquired immunity to reinfection (Olveda *et al.* 1996). In those under 14 years of age, the time to infection was the same for those with prior infections and those without prior infections. But for individuals aged 14–25 years, those with prior infections took significantly longer to become reinfected compared with those without prior infections. This noteworthy finding has also been found in other schistosome species (Dunne *et al.* 1992; Hagan *et al.* 1991). At the same time, transmigrant studies of schistosomiasis-naïve populations demonstrated that post-pubertal young adults developed partial resistance to reinfection much faster than children and pre-pubertal adolescents (Gryseels *et al.* 1994; Fulford *et al.* 1998).

Humans from endemic regions display a characteristic left-skewed convex distribution by age of prevalence and intensity of infection for the three major schistosome species. This suggests that the slow attrition of parasites from earlier infections is accompanied by a slowly acquired resistance to new infections. Other possible interpretations of the distribution data include slow parasite attrition coupled with reduced exposure, reduced worm fecundity in older hosts and non-immunologic factors such as increased host skin thickness with age. These confounders preclude the definitive demonstration of acquired human resistance from simple age and prevalence data, despite extensive iterations of the basic observation (Butterworth *et al.* 1989). To overcome these design limitations, several investigators have analyzed the intensity of reinfection following chemotherapeutic cure in groups of individuals with quantified water contact in schistosome endemic regions (Dunne *et al.* 1992; Hagan *et al.* 1991). Following chemotherapy, reacquisition and intensity of infection are strongly correlated with age. Young children rapidly become heavily reinfected while older children and adults become, if at all, lightly and slowly reinfected (Butterworth *et al.* 1988). Attenuated reinfection in older individuals cannot be attributed to reduced exposure; in one *S. mansoni* endemic community, intensities of infection peak in children aged 8–12 years, while water contact does not peak until 16–24 years (Butterworth *et al.* 1988). In a Ugandan community where adults have dramatically higher exposure than children due to fishing activities, children had higher reinfection intensities despite lower exposure (Kabatereine *et al.* 1999). These findings from all three schistosome species suggest that prior infection results in some acquired, partially protective immune responses.

The immunologic mechanisms underlying this age-dependent acquisition of partial resistance to schistosomiasis reinfection have not been identified. Informed speculations have begun to focus on pubertal events that might enhance immunity in schistosomiasis (Fulford *et al.* 1998). Recent evidence from malaria in Kenya provides clear evidence for the role of puberty in reducing rates of reinfection and degree of parasitemia (Kurtis *et al.* 2001). These observations and ideas have led us to propose a recently begun study of the influence of puberty on the development of immunity to *S. japonicum* in Leyte. More detail on this new project is provided in a later section.

S. japonicum-*associated morbidity*

Morbidity measures such as liver and spleen enlargement and degree of liver fibrosis were present in adults and adolescents although weakly related cross-sectionally to intensity of *S. japonicum* infection (Olveda *et al.* 1996; Wiest *et al.* 1993, 1994). Hepatomegaly and liver fibrosis were reversible with praziquantel treatment, especially in youth (McGarvey *et al.* 1995; Wiest *et al.* 1994). The eight-year longitudinal study in the Philippines indicated a remarkable rebound in hepatomegaly when intervals of treatment exceed one year after reinfection (Olveda *et al.* 1996; Olds *et al.* 1996).

A series of studies was begun to assess the impact of *S. japonicum* infection on childhood and adolescent health and function, reviewed in McGarvey (2000). This effort was an explicit attempt to learn if this infection exerted significant illness on children, since little research had been designed to measure 'subtle morbidity' apart from the classic liver and spleen enlargement. Childhood and adolescent malnutrition and reduced body size, muscle mass, and fatness adjusted for age were causally related in a dose-response manner to schistosomiasis japonica in two cross-sectional studies, one among over 1500 boys and girls aged 4–19 years in the endemic region of Leyte, and the other in over 230 children and adolescents from Jishan Island in Jiangxi Province (McGarvey *et al.* 1992, 1993). The greatest age-specific size and nutritional status differences between infected and non-infected individuals occurred during adolescence, age 16–17 for males and 12–13 for females. For example, in the Philippine study infected males were 7.8 cm shorter and 5.8 kg lighter than their uninfected counterparts.

Because the cross-sectional findings contain potential confounding by unmeasured factors such as poverty, diet, and non-helminthic infections on nutritional status, a randomized study of schistosomiasis japonica and nutritional status was performed among 170 *S. japonicum*-infected males and females, of mean age 11.4 years (McGarvey *et al.* 1996). Participants were randomized to praziquantel or placebo and followed-up six months after randomization for changes in fatness, assessed by the sum of skinfolds, and haemoglobin. Height, weight, and hookworm, *Trichuris* and *Ascaris* infection intensities were also measured at baseline and six-month follow-up. Six months after randomization the praziquantel group had significantly higher haemoglobin levels and sum of skinfolds than the placebo group, after adjustment for age, baseline values, and changes in the geoehelminth infections. These results show that *S. japonicum* infection is causally related to decreased nutritional status in children and adolescents, and suggests that chronic infection is partially responsible for the malnutrition and growth for age retardation observed in the cross-sectional studies.

Lastly, the effects of annual case-finding and treatment of *S. japonicum* infection on growth for age show dramatic and significant increases and decreases with changes in infection status (Olds *et al.* 1996). Infected males aged 8–12 years grew 4.8 cm and gained 2.6 kg in one year following parasitologic cure, compared with gains of 3.6 cm in stature and 2.1 kg in weight among those who were and remained uninfected. Those who became infected in a one-year interval grew 3.1 cm and gained 1.3 kg of weight. Because all infected individuals are treated, it is not possible to contrast the one-year growth of those who were and remained infected.

In addition to the evidence on the causal influence of schistosomiasis on nutritional status, an unpublished thesis indicated that prior nutritional status was predictive of later incidence of *S. japonicum* reinfection (Renza 1996). Nutritional status was assessed in 673 children and adolescents in the Leyte villages by the standard anthropometric indices: weight/height, height/age, weight/age, and sum of skinfolds/age. The risk of reinfection was adjusted for the prior year's infection intensity. Undernourished 12–24-year-olds were significantly twice as likely to become reinfected in one year and 1.5 times as likely over two years, relative to those who were adequately nourished. Age-stratified analyses showed that 12–15-year-olds had the greatest risk of one-and two-year reinfection. Undernutrition-mediated immune suppression may be partially responsible for these findings, as well as potential confounding by differential water exposure.

As part of the effort to determine the impact of *S. japonicum* on child health, a study of cognitive function was performed, using a double blind placebo controlled treatment trial conducted in Sichuan, China with 181 schoolchildren aged 5–16 years (Nokes *et al.* 1999). Three months after praziquantel treatment, there were significant age group by praziquantel treatment interaction effects in three of the five cognitive tests of working memory: Fluency, Picture Search and Free Recall. The effect of treatment was strongest in the youngest children aged 5–7 years. In the youngest children there was a significant positive main effect of treatment on Fluency, after controlling for sex, anthropometric, parasitic, and iron status. Thus, *S. japonicum* infection appears to be one cause of decreased performance of working memory in these younger schoolchildren. The presence of significant interactions between treatment and height-for-age, treatment and iron status, and treatment by *S. japonicum* intensity indicated that the most nutritionally vulnerable were the most likely to benefit in terms of improved performance on tests of working memory from the treatment.

The following sections present the background and brief description of two new projects that build upon previous research. The first pertains to our new study of the ecology and transmission of *S. japonicum*. The second builds on the observations about the role of age and puberty on the development of partial immunity.

Ecology and transmission of *S. japonicum* in the Philippines

Schistosoma japonicum is unique among the major schistosomes infecting humans in that zoonotic transmission is considered important, with domesticated animals serving as reservoir hosts of the parasite, with an amphibious snail as an inter-mediate host (Jordan *et al.* 1993). Although human infection and disease from *S. japonicum* have been reduced in China and the Philippines, as described above, further reductions in these endemic areas may be difficult because of the potential continual transmission from infected animals.

S. japonicum *infection in non-human animals*

Among the schistosomes infecting humans, *S. japonicum* has the widest variety of mammalian hosts, infecting over 40 domestic and wild animals (Pesigan *et al.* 1958; Cheng 1971; Carney *et al.* 1978; Kumar and Bubare 1986; He 1993). It is the only schistosome for which zoonotic transmission is considered important, and unique among helminthic zoonoses in that infections are transmitted naturally between man and other mammals with the infection maintained by all species (Nelson 1975).

Important studies on the biology of schistosomes which infect humans have been conducted using laboratory rodents and primates as experimental hosts. However, the validity and relevance of the results have been questioned by the responsible researchers themselves, since these animals are not natural hosts of the parasites (Basch 1991; Cheever *et al.* 1994). Only a few *S. japonicum* studies of natural definitive host–parasite relationships have been published despite the fact that many mammalian species serve as reservoir hosts of this zoonotic trematode (Wang 1959; Ho 1963; He 1993; Willingham *et al.* 1998). In a review of urgent needs for schistosome research, Hagan and Gryseels (1994) recommended that more research should be done on *S. japonicum* as it is the least well known of the schistosomes infecting humans, and in particular studies on host–parasite interactions in animals and their epidemiology in natural systems are needed.

Animals, including cattle, water buffalo, pigs, goats, dogs, cats, and wild rats, have been found to play an active role in contaminating the environment with *S. japonicum* eggs in the endemic countries of the Philippines, China and Indonesia (Mao 1948; Lung *et al.* 1958; Maegraith 1958; Pesigan *et al.* 1958; Dumag *et al.* 1981; Fernandez *et al.* 1982; Zheng *et al.* 1997; Wu *et al.* 1992; Chen, 1993; Brindley *et al.* 1995; Wang *et al.* 1998; McGarvey *et al.* 1999). Studies of the spatial distribution of animal faeces in endemic areas of China suggest that cattle dung contributes substantially to potential transmission of *S. japonicum* in that country (He *et al.* 1991; Wu *et al.* 1992; McGarvey *et al.* 1999). Similar studies in the Philippines have not been conducted for over 20 years and no systematic estimates exist of the contribution of domestic animals to *S. japonicum* transmission among humans in either country. In view of this, combined human and veterinary epidemiological studies are urgently needed to improve understanding of the zoonotic implications of schistosomiasis japonica and to create a basis for suitable control measures in domestic animals.

Older studies from the Philippines do provide some indications of the relative contributions of specific animals to *S. japonicum* transmissions, as well as recent unpublished studies on carabaos. Table 7.1 presents a summary of findings from one of the older studies (Dumag *et al.* 1981) based on surveys in 50 barangays (villages) in Dagami, Leyte, the Philippines in 1979.

Because of the remarkably high percentage of cercariae from field rats in this 1979 Leyte study, it appears that they make a large contribution to transmission, despite their lower egg hatchability. Similar data from China are not available for all these animals, but as described above it appears from study of faecal distributions

Table 7.1 Contribution of definitive hosts to transmission in 50 villages in Dagami, Leyte in 1979

Parameters	Dogs	Pigs	Carabao	Rats	Humans
Est total pop	1756	2672	1424	1,408,806	20,121
Prevalence % in sample	7.7%	4.2%	0.07%	73.7%	18.5%
Mean epg/d	1747.8	1367.2	9513.7	12.7	11.2
Hatchability %	19.5%	30.8%	29.6%	10.7%	42.4%
Est. pop inf.	135	112	1	1,038,572	3723
Est # cerc/ind-day	339.9	421.6	2814.1	1.36	4.8
Est % Cerc/Total	3.0%	3.1%	0.02%	92.6%	1.2%

Source: adapted from Table 1 in Dumag *et al.* (1981).

in fields that carabao contribute the majority of *S. japonicum* eggs and thus appear to have the largest influence on transmission (He *et al.* 1991; Wu *et al.* 1992). These findings indicate there are different transmission patterns/roles for animal reservoir hosts in the two countries.

The fact that animal reservoir hosts are considered to play an integral role in transmission of *S. japonicum* to each other and humans has led to new strategies for control of *S. japonicum* infection in humans. Basic and operational research is being planned, directed at chemotherapy and vaccination of animals to eliminate the reservoir for human transmission and reduce transmission (Shi *et al.* 1998; McManus *et al.* 1998; Nara *et al.* 1998; Zhou *et al.* 1998; Wang *et al.* 1998; many of these described in McGarvey *et al.* 1999). However, as mentioned, there is very little baseline information on which to base these strategies. In addition, based on available Philippine data, field rats appear to contribute quite a lot to transmission. Thus, such domestic animal vaccines as those planned for China may not be applicable in the Philippines.

Oncomelania snails and S. japonicum *transmission*

The ecology of Oncomelania snails must be understood to account for the dynamics of *S. japonicum* transmission. Snails are the common source of environmental exposure that delivers the daily infective dose of cercaria to both human and non-human hosts. The proportion of snails infected with schistosomes at any given time depends upon a complex interaction of factors: distribution and behavior of definitive hosts; relative susceptibility of the snail to infection; and temperature and rainfall. In the Philippines, where temperatures are fairly constant and rainfall occurs throughout the year, reproduction of *O. quadrasi* takes place continuously, while in areas where there is a definite dry season, these snails do not occur (Pesigan *et al.* 1958). Snail information is mandatory to measure exposure rather than just water contact. Geographic variation in snail infection exists, whether over the entire endemic regions or a microspace (Aligui 1997). Exposure is usually ascertained by cercariometry but such techniques are extremely intensive, subject to sampling bias, and not always necessary for understanding the risk of reinfection to definitive hosts (Aligui 1997).

The distribution of snails is a key aspect in the study of exposure. Snail distribution has been found to be uneven (Makiya *et al.* 1981). This was confirmed by Aligui *et al.* (1997) in the water contact studies in Leyte. The distribution is affected by migration of snails and the increase in water level. These factors imply that the ecological changes in bionomics and limnology are closely related to the infection potential of endemic areas and show the importance of assessing the snail infection over the period of water contact observation in parallel with the ecological characterization of the endemic foci.

Snail control may be achieved principally through chemical, environmental, or biological means (Madsen and Christensen 1992). The focal use of molluscicides at epidemiological important water contact sites is possible but requires repeated long-term application by highly skilled personnel. Habitat modification through drainage, land reclamation and good water management practices may be an effective means of eliminating the snail host of *S. japonicum*. In the Philippines (Santos 1984; Makiya *et al.* 1981) and in China (Mao and Shao 1982), a wide range of environmental modifications has been employed with a great reduction of prevalence of schistosomiasis. A number of environmental control measures exist, including stream channelization, seepage control, canal lining, canal relocation with deep burial of snails, proper drainage in irrigation schemes, removal of vegetation, earth filling, ponding, and improved agricultural practices, which could be used to reduce snail population densities, especially in irrigation schemes. Amphibious Oncomelania snails are particularly sensitive to such measures, owing to the nature of their habits (Cheng 1971; Mao and Shao 1982).

At a meeting in Wuxi, Jiangsu, China, on 8–10 September 1998 many of these studies were reviewed and a consensus reached about the need for more research on the role of animals in transmission of *S. japonicum* (McGarvey *et al.* 1999). The current epidemiologic situation of *S. japonicum* infection in humans and animals was described. The putative role of *S. japonicum* infection in animals on transmission in humans was discussed. The need for development of animal vaccines against *S. japonicum* was also described and discussed. Emphasis was placed on promoting an inter-sectoral approach to research, surveillance, and control, and on facilitating interregional cooperation and collaboration.

One year after that meeting, the US National Institutes of Health announced a request for applications on the topic of the ecology of infectious diseases and we were fortunate to receive research support to formally investigate the ecology and transmission of *S. japonicum* in the Philippines. The overall objective of this new research is the development of a dynamic model of the influence of anthropogenic changes due to rice farming on the transmission of *Schistosoma japonicum* in the endemic areas of the Philippines. The model will include several species, their population biology and behavior, including humans, animal reservoir hosts and snails. Rice farming, its spatial and temporal variation and its multiple impacts on physical, hydrologic and animal determinants of human *S. japonicum* infection, is defined as anthropogenic environmental change. The model will allow the prediction of the potential effects of anthropogenic environmental changes stemming from aspects of rice farming over time and space on the transmission of *S. japonicum* to humans. Rice farming factors will include: variation in rice farming practices, agricultural expansion and intensification; domesticated animal numbers and level of infection, transmission behaviors and management practices; water use and management practices. Developing generalizable predictive ecological models of schistosomiasis transmission to humans is a fundamental component of the ongoing effort to develop cost-effective and acceptable disease control strategies. There are at present no published population dynamic models for *S. japonicum*, which remains endemic in the Philippines and China despite 20 years of treatment of infected humans and some treatment of animals in China. Fieldwork for this project has commenced and an early version of a transmission model has been developed using the data from the 1980s and 1990s. This model will then be refined with the newly collected information, and in particular by measuring the intensity of infection simultaneously in all mammal species and snails. Several villages will be sampled to ensure generalizability of the results.

One of the challenges for this project is accurate determination of human and animal exposure to potentially infective water, which could act as a confounding factor in detecting the effect of water management on infection. The work will be starting with a recently developed index of exposure based on snail infection prevalence at specific water sites, and detailed sampling of observed human–water contact including time of day, and degree and duration of contact (Aligui 1997). This will be supplemented by a validated water contact questionnaire due to the logistic ease for studying many villages and individuals.

Puberty, malnutrition and immunity to *S. japonicum*

Building on the group's observations about the impact of schistosomiais on malnutrition and growth perturbations during adolescence, the evidence for an age-related development of immunity to reinfection, and specific observations about the impact of puberty on the development of immunity to malaria, a new study has been started. The overall objective of this research is to prospectively investigate the interrelationships among puberty, protective immune responses, and nutritional status in adolescent Philippine residents of a *Schistosoma japonicum* endemic area. The rationale is that determining immunologic and developmental predictors of resistance in naturally exposed humans should be a fundamental component of the effort to develop vaccines against schistosomiasis japonica.

One interpretation of the classic negative age and intensity of infection relationship is that years of cumulative exposure are required for expression of resistance. A recent alternative hypothesis is that the hormonal changes of puberty, e.g. increasing dehydroepiandrosterone (DHEA) and its sulphate (DHEAS), collectively referred to as DHEA(S), not cumulative exposure, initiate and promote the development of protective immune responses to schistosome infections (Fulford *et al.* 1998). This has been demonstrated for malaria (Kurtis *et al.* 2001).

Schistosomiasis is causally linked to malnutrition, leading to the hypothesis that chronic infection results in attenuated growth for age and delayed pubertal development (McGarvey 2000). Animal models of chronic parasitemia have identified TNF-alpha and IL-6 as mediators of malnutrition and cachexia. Production of these mediators is significantly attenuated by increasing DHEA(S) levels. These interrelationships suggest that a DHEA(S)-modulated cycle of infection, pubertal delay, and malnutrition may be responsible for the marginal nutritional status of schistosome-infected adolescents.

The 18-month longitudinal interrelationships among specific antibody responses to specific schistosome antigens and rates of reinfection and intensity of reinfection will be studied. Longitudinally protective immune responses are expected to be found. It is hypothesized that DHEAS and pubertal assessments will moderate these relationships between antibody responses and reinfection. Namely, those adolescents further along in puberty or with higher DHEAS levels will have an increased expression of the protective immune responses. Furthermore, anthropometric and serum measures of malnutrition will be associated with circulating mediators of inflammation such as TNF-alpha or cachexia. These variables will be directly related to delays in puberty and DHEAS levels, thus further slowing the development of protective immune responses. Because of the obvious ethical issues, adolescents will not be followed throughout adolescence and after 18 months their schistosomiasis will be treated. But more will have been learned about the potential role of pubertal and nutritional status and inflammatory processes in the development of partial immunity to *S. japonicum* infection, and their complex interactions.

Conclusions

It is a truism in human population biology and health research that interdisciplinary perspectives will yield findings and insights greater than available within the individual disciplines. This truism is repeated here both as an admonition and a promise. The authors of this contribution include the fields of biological anthropology, paediatrics, human epidemiology, health economics, immunology, human pathology, veterinary medicine, veterinary epidemiology, and internal medicine. In addition we share the general perspectives of naturalists, field workers, laboratory scientists and evolutionists. This has allowed us to start two innovative projects that should provide answers to specified hypotheses about the transmission and immunology of *S. japonicum*. They should also provide new questions!

References

Aligui, G.L. (1997) 'Quantifying human *Schistosoma japonicum* infection and exposure: errors in field diagnosis and uncertainties in exposure measurements', PhD Thesis, Brown University, RI.

Aligui, G.L., Mor, V., McGarvey, S.T., Olds, G.R. and Olveda, R.M. (1997) 'A proposed exposure model specific for *Schistosoma japonicum*', *Am J Trop Med Hyg* (Supplement), 57: 241.

Basch, P.F. (1991) *Schistosomes: Development, Reproduction and Host Relations*, New York: Oxford University Press, pp. 67–75.

Bergquist, R., Malone, J.B. and Kristensen, T.K. (2000) 'Health maps: a global network on schistosomiasis information systems and control of snail-borne diseases', *Parasitol Today*, 16: 363–4.

Blas, B.L., Tormis, L.G. and Portillo, L.A. (1988) 'The schistosomiasis control component of the national system improvement project in Leyte, Philippines. Part III. An attempt to study the economic loss arising from *Schistosoma japonicum* infection', Proceedings of SEAMEO-TROPMED seminar, 13–16 June 1988, Surat Thani, Thailand.

Brindley, P.J., Ramirez, B., Tiu, W., Wu, G., Wu, H.W. and Yi, X. (1995) 'Networking schistosomiasis japonica', *Parasitol Today*, 11: 163–5.

Butterworth, A.E., Fulford, A.J.C., Dunne, D.W., Ouma, J.H. and Sturrock, R.F. (1988) 'Longitudinal studies in human schistosomiasis', *Philosophical Transactions of the Royal Society of London Series B: Biological Sciences*, 321: 495–500.

Nara, T., Chen, H. and Hirayama, K. (1998) 'Vaccination trial of domestic pigs with recombinant paramyosin against *S. japonicum* in China – a strategy for controlling the parasite in reservoir livestock', in Proceedings of International Symposium on Schistosomiasis: The Epidemiology and Host–Parasite Relationships of *Schistosoma japonicum* in Definitive Hosts, 8–10 September 1998, Jiangsu Institute of Parasitic Diseases, Wuxi, Jiangsu, China, p. 61.

Nelson, G.S. (1975) 'Schistosomiasis', in W.T. Hubbert, W.F. McCulloch and P.R. Schnurrenberger (eds) *Diseases Transmitted from Animals to Man*, pp. 620–40, Springfield, IL: Charles C. Thomas.

Nokes, C., McGarvey, S.T., Shiue, L., Wu, G., Wu, H., Bundy, D.A.P. and Olds, G.R. (1999) 'Evidence for an improvement in cognitive function following treatment of *Schistosoma japonicum* infection in Chinese primary schoolchildren', *Am J Trop Med Hyg*, 60: 556–65.

Olds, G.R., Olveda, R., Wu, G., Wiest, P.M., McGarvey, S.T., Aligui, G. and Zhang, S. (1996) 'Immunity and morbidity in *Schistosomiasis japonicum* infection', *Am J Trop Med Hyg*, 55: 121–6.

Olveda, R.M., Fevidal, P., Veyre, F.D., Icatalo, F.C. and Domingo, E.O. (1983) 'Relationship of prevalence and intensity of infection to morbidity in schistosomiasis japonica: a study of three communities in Leyte, Philippines', *Am J Trop Med Hyg*, 32: 1312–21.

Olveda, R.M., Daniel, B.L., Ramirez, B.D., Aligui, G.D., Acosta, L.P., Fevidal, P., Tiu, E., de Veyra, F., Peters, P.A., Romulo, R., Domingo, E., Wiest, P.M. and Olds, G.R. (1996) 'Schistosomiasis japonica in the Philippines: the long-term impact of population-based chemotherapy on infection, transmission, and morbidity', *Journal of Infectious Diseases*, 174: 163–70.

Pesigan, T.P., Farooq, M., Hairston, N.G., Jauregui, J.J., Garcia, E.G., Santos, A.T. and Besa, A.A. (1958) 'Studies on *Schistosoma japonicum* infection in the Philippines. 3. Preliminary control experiments', *Bull WHO*, 19: 223–61.

Peters, P.A.S. and Kazura, J.W. (1987) 'Update on diagnostic methods for Schistosomiasis', *Balliere's Clin Trop Med Communic Dis*, 2: 419–33.

Renza, E. (1996) 'Undernutrition and reinfection with schistosomiasis', Brown University.

Santos, A.T. Jr. (1984) 'Prevalence and distribution of schistosomiasis in the Philippines: a review', *Southeast Asian Journal of Tropical Medicine and Public Health*, 7: 133–6.

Shi, F., Shen, W. and Lin, J. (1998) 'The control of schistosomiais japonica in domestic animals in China', in Proceedings of International Symposium on Schistosomiasis: The Epidemiology and Host–Parasite Relationships of *Schistosoma japonicum* in Definitive Hosts, 8–10 September 1998, Jiangsu Institute of Parasitic Diseases, Wuxi, Jiangsu, China, p. 36.

Sleigh, A., Li, X., Jackson, S. and Huang, K. (1998a) 'Eradication of schistosomiasis in Guangxi, China. Part 1: Setting, strategies, operations and outcomes, 1953–92', *Bull WHO*, 76: 361–72.

Sleigh, A., Jackson, S., Li, X. and Huang, K. (1998b) 'Eradication of schistosomiasis in Guangxi, China. Part 2: Political economy, management strategy and costs, 1953–92', *Bull WHO* 76: 497–508.

Sleigh, A., Li, X., Jackson, S. and Huang, K. (1998c) 'Eradication of schistosomiasis in Guangxi, China. Part 3: Community diagnosis of the worst-affected areas and maintenance strategies for the future'. *Bull WHO*, 76: 581–90.

Wang, H.Z., Yuan, H.C. and Feng, Z. (1998) *Epidemic Status of Schistosomiasis in China*, Shanghai: Ministry of Health, People's Republic of China.

Wang, X.Y. (1959) *Schistosomiasis in Domestic Animals*, Shanghai: Shanghai Publishing House for Sciences and Technology (in Chinese).

Wiest,, P.M., Wu, G., Zhang, S., Yuan, J., Peters, P., McGarvey, S.T., Tso, M., Olveda, R. and Olds, G.R. (1992) 'Morbidity due to schistosomiasis japonica in the People's Republic of China', *Trans Roy Soc Trop Med Hyg*, 86: 47–50.

Wiest, P.M., Wu, G., Zhang, S., McGarvey, S.T., Tan, E., Yuan, J., Peters, P., Olveda, R. and Olds, G.R. (1993) ' Schistosomiasis japonica on Jishan Island, Jiangxi Province People's Republic of China: persistence of hepatic fibrosis after reduction of the prevalence of infection with age', *Trans Roy Soc Trop Med Hyg*, 87: 290–4.

Wiest, P.M., Wu, G., Zhong, S., McGarvey, S.T., Yuan, J., Olveda, R.M., Peters, P.A.S. and Olds, G.R. (1994) ' Impact of annual screening and chemotherapy with praziquantel on schistosomiasis japonica on Jishan Island, People's Republic of China', *Am J Trop Med Hyg*, 51: 162–9.

Willingham, A.L. III, Hurst, M., Bøgh, H.O., Johansen, M.V., Lindberg, R., Christensen, N.Ø. and Nansen, P. (1998) '*Schistosoma japonicum* in the pig: the host–parasite relationship as influenced by the intensity and duration of experimental infection', *Am J Trop Med*, 58: 248–56.

World Health Organization (1993) *The Control of Schistosomiasis*, 2nd Report of the WHO Expert Committee (Technical Report Series 830), Geneva: WHO.

Wu, Z., Liu, Z.D., Pu, K.M., Hu, G.H., Zhou, S.J., Zhou, S.Y., Zhang, S.J. and Yuan, H.C. (1992) 'Role of human and domestic animal reservoirs of schistosomiasis japonica in Dongting and Boyang Lake regions', *Chinese Journal of Parasitology and Parasitic Diseases*, 10: 194–7 (in Chinese with English abstract).

Yuan, H., Jiagang, G., Bergquist, R., Tanner, M., Xianyi, C. and Huanzeng, W. (2000) 'The 1992–99 World Bank Schistosomiasis Research Initiative in China: outcomes and perspectives', *Parasitol Int*, 49: 195–207.

Zheng, J., Zheng, Q., Wang, X. and Hua, Z. (1997) 'Influence of livestock husbandry on schistosomiasis transmission in mountainous regions of Yunnan Province', *SE Asia J Trop Med Pub Hlth*, 28: 291–5.

Zhou, Y., Chen, Y. and Xie, M. (1998) 'Analysis of chemotherapy reducing the transmission and infection of domestic animals in the Dongting Lake region', in Proceedings of International Symposium on Schistosomiasis: The Epidemiology and Host–Parasite Relationships of *Schistosoma japonicum* in Definitive Hosts, 8–10 September 1998, Jiangsu Institute of Parasitic Diseases, Wuxi, Jiangsu, China, p. 67.

8 Unravelling gene–environment interactions in type 2 diabetes

Nicholas J. Wareham

Type 2 diabetes is an increasingly common disorder which gives rise to considerable morbidity and mortality (Marks 1996). It is the commonest cause of preventable blindness in adults of working age, a major cause of end-stage renal disease (Nelson *et al.* 1995) and lower limb amputation (The Lower Extremity Amputation Study Group 1995). It is, perhaps most importantly, a major risk factor for coronary heart disease, which occurs at least twice as commonly in people with diabetes (Fuller *et al.* 1983; Kannel *et al.* 1979). The pattern of the rise in the prevalence of type 2 diabetes around the world, particularly in previously undeveloped countries, provides support for the notion that this disorder results from an interaction between genetic and environmental factors. This chapter describes the evidence that suggests such interactions exist and debates the most appropriate study designs which could be employed to detect them, using evidence from several recently described examples.

What is the evidence that gene–environment interactions exist in type 2 diabetes?

The global variation in the prevalence of type 2 diabetes is marked. Figure 8.1 shows the prevalence of type 2 diabetes and impaired glucose tolerance in various countries around the world (King and Rewers 1993). These prevalence estimates have been obtained by adding the prevalence of clinically diagnosed diabetes to that of undiagnosed disease. This is necessary because up to half of all potentially detectable cases at any given time are undetected, as this disorder has a slow insidious onset and presents without acute metabolic disturbance (Harris 1993). Therefore clinically the distinction between normality and abnormality is blurred and a standardised test, the 75 g oral glucose tolerance test, has to be utilised to find the clinically undiagnosed cases. The geographical pattern of variation suggests that prevalence rates are lowest in rural areas of developing countries, are generally intermediate in developed countries, but are highest in certain ethnic groups who have adopted western lifestyle patterns (King and Rewers 1993). This is particularly apparent in the two populations with highest prevalence, the Nauruan people in Micronesia and the Pima Indians in Phoenix, Arizona. Both of these populations have undergone rapid changes in dietary and physical activity lifestyle and have

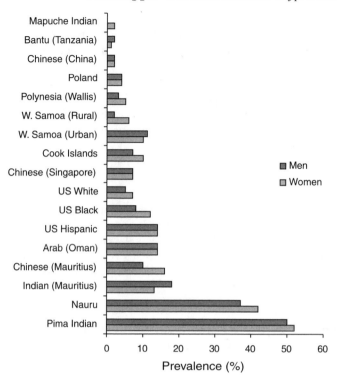

Figure 8.1 Age and sex standardised prevalence of diabetes in men and women aged 30–64 years.

demonstrated marked increases in the prevalence of obesity and diabetes (Zimmet *et al.* 1984; Knowler *et al.* 1990). In the 1960s a hypothesis was developed to explain the occurrence of diabetes susceptibility in these populations (Neel 1962). This 'thrifty genotype' hypothesis speculates that a genetic predisposition to obesity and diabetes would be advantageous in an evolutionary sense in times of food scarcity as it would promote efficient retention of energy stores, but would become disadvantageous in times of relative food abundance and low energy expenditure (Swinburn 1996). Populations that had survived periods of marked food scarcity in their development would, therefore, become enriched for a diabetes gene. This hypothesis is supported by the history of high prevalence populations. The Pima Indians, for example, survived migration down the ice-free corridor on the western side of the United States and can be traced archeologically as the tribe moved from one buffalo kill-site to another (Wendorf 1989). Such survival pressures would have also faced the peoples of the South Pacific, who would have been better adapted to ocean voyages if their metabolism was adapted to storing energy efficiently, particular as excess body fat (Houghton 1991).

The thrifty genotype hypothesis would predict that individuals from at-risk populations who migrate to countries where lifestyles are more westernised would be at increased risk of developing diabetes. This is demonstrably true, for example, among migrants from Japan to the United States (Fujimoto *et al.* 1987) and among people who migrate from Tokelau to New Zealand (Stanhope and Prior 1980). However, inferences from migration studies are limited as migrants may differ in other respects from static populations of the same ethnicity. Stronger evidence comes from cohort studies in which the importance of genetic factors for diabetes has been demonstrated by familial aggregation (Lillioja *et al.* 1987), concordance in twin studies (Tattersall and Pyke 1972; Barnett *et al.* 1981) and impact of partial or full Indian heritage on diabetes risk in high prevalence populations like the Pimas (Knowler *et al.* 1990). Similarly the importance of non-genetic risk factors such as obesity (Chan *et al.* 1994), physical inactivity (Manson *et al.* 1992), and dietary intake of fat (Marshall *et al.* 1991) has been demonstrated in population-based cohort studies in which the incidence of diabetes has been measured (Hamman 1992). The evidence that these genetic and environmental factors interact comes from studies suggesting that risk is higher in certain high risk groups given the same environment exposure. For example, people with a family history of diabetes are more at risk of developing diabetes than people without a family history, and that risk is modified by the presence of obesity and inactivity (Morris *et al.* 1988; Hanson *et al.* 1995; Sargeant *et al.* 2000).

Should studies of interaction treat diabetes as a binary state or as part of a continuum?

Given that gene–environment interactions are likely to exist, the next question is how best to try to identify them. The classical epidemiological methods include either cross-sectional case-control studies or case-control studies nested within a cohort. The cross-sectional case-control approach is relatively simple and for a disorder with a relatively high prevalence like diabetes, it is a fairly easy task to identify diagnosed cases within a population. This approach would be fine for a purely genetic study, but the examination of environmental influences is likely to be affected by recall bias since the assessment of diet and physical activity occurs after an individual is aware that they have diabetes. As provision of information and advice about physical activity and dietary change is central to the management of diabetes, assessment of these behaviours after the attribution of the disease label is likely to result in a biased indication of usual pre-disease status. The normal response in this situation is to design prospective studies in which diet is assessed before the occurrence of disease. This strategy is expensive as many individuals need to have their environmental factors assessed for the relatively few who progress to disease. Even though diabetes is a prevalent disorder, the absolute incidence is not high (6 per 1000 person years of follow-up in people aged 40–65 years (Wareham *et al.* 1999)) and if measurement of exposures is expensive, this design is inefficient as measurements will have been collected on 157 individuals who do not progress to diabetes for every one that does. The nested case-control approach is a solution to this efficiency problem as exposure data are collected but not analysed at baseline, and are then only processed and analysed at follow-up on those individuals who have become cases (plus a selection of controls). This combines the efficiency of a case-control study with the elimination of exposure recall bias seen in cohort studies. This design would be suitable for the study of gene–environment interaction in type 2 diabetes, but would require a large initial cohort and long-term follow-up in order to generate sufficient cases. In general a case-control study of interaction needs at least four times as many cases as a simple association study (Smith and Day 1984). Thus if something of the order of more than 5000 clinically incident cases needed to be collected, then a cohort would need to accrue several million person years of follow-up.

A completely different alternative strategy is to consider diabetes not as a categorical state, but as an extreme of a continuum. In this situation, analyses of association with genetic factors can be undertaken with those continuous variables or quantitative traits as the outcome rather than diabetes as a binary state. This approach has a number of attractions. Firstly the reduction of diabetes to a binary state may be somewhat artificial. In most populations the levels of fasting and post-glucose load glycaemia on which the diagnosis of diabetes is based are normally distributed (Williams *et al.* 1995) with no evidence of a clear distinction between those who are diseased and those who are not. Perhaps more importantly, unlike the risk of microvascular disease (McCance *et al.* 1994) the cardiovascular risk associated with levels of hyperglycaemia is linear without an obvious threshold (Khaw *et al.* 2001). Secondly, type 2 diabetes is a heterogeneous disorder, which results from either insulin resistance or beta cell dysfunction or a combination of these two processes (DeFronzo *et al.* 1992). Studying the impact of genetic and environmental influences on continuously distributed parameters, such as the fasting insulin concentration, which are related to insulin sensitivity (Laakso 1993), may provide clearer evidence of causal pathways underlying diabetes than studies of hyperglycaemia itself which is simply the final common shared phenotype. A third advantage of quantitative trait studies is that they can be undertaken in a cross-sectional manner because the issues of recall bias that are so typical of cross-sectional case-control studies in diabetes are not a problem for continuous traits.

Proven examples of gene–environment interaction in type 2 diabetes

To date, few examples have been reported of gene–environment interactions on continuous traits related to glucose and insulin metabolism. In an experimental study, Salas and colleagues demonstrated differential effects of manipulation of dietary fat on basal and post-glucose load glucose and insulin in individuals with a polymorphism in the Apo C-III gene (Salas *et al.* 1998). In a recent study Ukkola and colleagues demonstrated differential responses to over-feeding for individuals with polymorphisms in insulin-like growth factors 2 (IGF2) and IGF-binding protein 1 (IGFBP1) (Ukkola *et al.* 2001). Non-experimental studies include an analysis of the differential effects of alcohol on glycaemic control in patients with diabetes with active or inactive forms of alcohol dehydrogenase 2 (Murata *et al.* 2000) and two recent publications from the Ely Study. In an analysis of the effect of common polymorphisms in the β-adrenoceptor (βAR-2), Meirhaeghe and colleagues demonstrated that the effect of the Gly16Arg polymorphism was modulated by the levels of habitual physical activity (Meirhaeghe *et al.* 2001). This analysis was focused on activity as Meirhaeghe and colleagues had previously demonstrated an interaction between this polymorphism and reported physical activity on levels of obesity (Meirhaeghe *et al.* 1999). A second example of interaction in this study involved a common polymorphism in the gene for the peroxisome proliferator-activated receptor γ (PPARγ). This receptor is known to be important in determining insulin sensitivity, as a new class of drugs, the thiazolidinediones, act through this receptor to improve insulin resistance and individuals with loss of function mutations have severe insulin-resistant diabetes (Barroso *et al.* 1999). In a meta-analysis of case-control studies, a common polymorphism in the PPARγ gene (Pro12Ala) was associated with reduced diabetes risk (Altshuler *et al.* 2000). The search for gene–environment interaction is likely to result in fewer false positive findings if it is based on an understanding of biology rather than a random search for statistical interaction. In this example Luan *et al.* postulated *a priori* that an interaction could exist with the type of fat in the diet because previous work had suggested that PPARγ may be a fatty acid sensor with the affinity of fatty acids for the receptor varying according to their chain length and degree of desaturation (Kliewer *et al.* 1997; Desvergne and Wahli 1999). The observational data in the Ely Study demonstrated an interaction between the Pro12Ala polymorphism with the ratio of polyunsaturated to saturated fats in the diet in determining insulin sensitivity (Luan *et al.* 2001a).

Size of study required to detect gene–environment interaction

If quantitative trait analysis is to be undertaken, the size and therefore power of these studies to detect gene–environment interaction is a major issue. Although existing formulae exist for case-control studies where the outcome is reduced to a binary state, until recently formulae had not been produced for studies where both the exposure of interest and the outcome were continuously distributed. In considering this situation, Luan and colleagues reported how power is dependent upon allele frequency, the size of the association between the exposure and the outcome, and the strength of the interaction term (Luan *et al.* 2001b). In an extension to this work, they have analysed the extent to which the ability to detect gene–environment interaction in a given situation is dependent upon measurement error in the assessment of the outcome and the exposure. This analysis clearly indicates that when poor measures of exposure are utilised, sample sizes have to be very large indeed to detect gene–environment interaction on continuous traits. For example, if the goal were to identify a genetic effect on insulin sensitivity that was modified by physical activity, then a study of more than 150,000 people would be required to detect a doubling of effect size for those with the minor allele of population frequency of 20 per cent if proxy measures of exposure and outcome were used. Even if better measures of outcome such as a single measure of fasting insulin were available, then the study would still need between 30,000 to 50,000 individuals. However, if a study not only employed a better measure of outcome but also added a reasonable measure of exposure, for example an objective quantitative measure of activity from heart rate monitoring (Wareham *et al.* 2000) rather than a questionnaire, then the same interaction could be detected in 5000 individuals. The magnitude of the difference in sample size between the alternative study designs is a further illustration of the importance in certain epidemiological studies of aiming for better measurement rather than simply for more subjects (Phillips and Davey Smith 1993).

Conclusions

The descriptive epidemiology of type 2 diabetes and findings from cohort studies suggest that this disorder originates in large part from a complex interaction between genetic and environmental factors. Determining the detail of these interactions using the nested case-control design may be optimal, but is a long term and expensive strategy. Quicker and cheaper results may be obtained by studying interaction on the quantitative traits that underlie diabetes. However, the power of such studies to detect interaction is highly dependent on the precision with which non-genetic exposures are measured. Unravelling these interactions will undoubtedly shed light on the aetiology of diabetes and will, it is hoped, lead to opportunities for targeted prevention. Recent studies in high risk groups such as people with impaired glucose tolerance suggest that the incidence of diabetes can be reduced by more than 50 per cent by interventions aimed at changing dietary and physical activity behaviour (Pan *et al.* 1997; Tuomilehto *et al.* 2001). However, it may be that individuals with a particular genotype are particularly susceptible to the negative metabolic consequences of sedentary living, and that they conversely, therefore, would have most to gain from a targeted preventive intervention programme. Understanding how to detect these individuals and which environmental factors a programme should attempt to manipulate is a major goal of studies that attempt to unravel gene–environment interaction.

References

Altshuler, D., Hirschorn, J.N., Klannemark, M., Lindgren, C.M., Vohl, M.-C., Nemesh, J., Lane, C.R., Schaffner, S.F., Bolk, S., Brewer, C., Tuomi, T., Gaudet, D., Hudson, T.J., Daly, M., Groop, L. and Lander, E.S. (2000) 'The Common PPARγPro12Ala polymorphism is associated with decreased risk of type 2 diabetes', *Nature Genetics*, 26: 76–80.

Barnett, A.H., Eff, C., Leslie, R.D.G. and Pyke, D.A. (1981) 'Diabetes in identical twins', *Diabetologia*, 20: 87–93.

Barroso, I., Gurnell, M., Crowley, V.E.F., Agostini, M., Schwabe, J.W., Soos, M.A., Li Maslen, G., Williams, T.D.M., Lewis, H., Schafer, A.J., Chatterjee, V.K.K. and O'Rahilly, S.O. (1999) 'Dominant negative mutations in human PPARγ associated with severe insulin resistance, diabetes mellitus and hypertension', *Nature*, 402: 23–30.

Chan, J.M., Rimm, E.B., Colditz, G.A., Stampfer, M.J. and Willett, W.C. (1994) 'Obesity, fat distribution, and weight gain as risk factors for clinical diabetes in men', *Diabetes Care*, 17: 961–9.

DeFronzo, R.A., Bonadonna, R.C. and Ferrannini, E. (1992) 'Pathogenesis of NIDDM: a balanced overview', *Diabetes Care*, 15: 318–68.

Desvergne, B. and Wahli, W. (1999) 'Peroxisome proliferator-activated receptors: nuclear control of metabolism', *Endocr Review*, 20: 649–88.

Fujimoto, W.Y., Leonetti, D.L., Kinyour, J.L. *et al.* (1987) 'Prevalence of diabetes mellitus and impaired glucose tolerance among second generation Japanese-American men', *Diabetes*, 36: 721–9.

Fuller, J.H., Shipley, M.J., Rose, G., Jarrett, R.J. and Keen, H. (1983) 'Mortality from coronary heart disease and stroke in relation to degree of glycaemia: the Whitehall study', *British Medical Journal*, 287: 867–70.

Hamman, R.F. (1992) 'Genetic and environmental determinants of non-insulin-dependent diabetes mellitus (NIDDM)', *Diabetes Metabolism Rev*, 8: 287–338.

Hanson, R.L., Pettit, D.J., Bennett, P.H., Narayan, K.M.V., Fernandes, R., de Courten, M. and Knowler, W.C. (1995) 'Familial relationships between obesity and NIDDM', *Diabetes*, 44: 418–22.

Harris, M.I. (1993) 'Undiagnosed NIDDM: clinical and public health issues', *Diabetes Care*, 16: 642–652.

Houghton, P. (1991) 'Selective influences an morphological variation amongst Pacific homo sapiens', *Journal of Human Evolution*, 21: 41–51.

Kannel, W.B. and McGee, D.L. (1979) 'Diabetes and glucose tolerance as risk factors for cardiovascular disease: the Framingham Study', *Diabetes Care*, 2: 120–6.

Khaw, K.-T., Wareham, N.J., Luben, R., Bingham, S., Oakes, S. and Welch, A. (2001) 'Glycated haemoglobin, diabetes, and mortality in men in Norfolk cohort of European Prospective Investigation of Cancer and Nutrition (EPIC-Norfolk)', *British Medical Journal*, 322: 15–18.

King, H. and Rewers, M. (1993) 'Global estimates for prevalence of diabetes mellitus and impaired glucose tolerance in adults', *Diabetes Care*, 16: 157–77.

Kliewer, S.A., Sundseth, S.S., Jones, S.A., Brown, P.J., Wisely, G.B., Koble, C.S., Devchand, P., Wahli, W., Willson, T.M., Lenhard, J.M. and Lehmann, J.M. (1997) 'Fatty acids and eicosanoids regulate gene expression through direct interactions with peroxisome proliferator-activated receptors alpha and gamma', *Proc Natl Acad Sci USA*, 94: 4318–23.

Knowler, W.C., Pettit, D.J., Saad, M.F. and Bennett, P.H. (1990) 'Diabetes mellitus in the Pima Indians: incidence, risk factors and pathogenesis', *Diabetes Metabolism Review*, 6: 1–27.

Laakso, M. (1993) 'How good a marker is insulin level for insulin resistance?', *American Journal of Epidemiology*, 137: 959–965.

Lillioja, S., Mott, D.M., Zawadzki, J.K., Young, A.A., Abbott, W.G., Knowler, W.C., Bennett, P.H., Moll, P. and Bogardus, C. (1987) 'In vivo insulin action is a familial characteristic in nondiabetic Pima Indians', *Diabetes*, 36: 1329–35.

Luan, J., Browne, P.O., Harding, A.-H., Halsall, D.J., O'Rahilly, S., Chatterjee, V.K.K. and Wareham, N.J. (2001a) 'Evidence for gene-nutrient interaction at the PPAR? locus', *Diabetes*, 50: 686–9.

Luan, J.A., Wong, M.Y., Day, N.E. and Wareham, N.J. (2001b) 'Sample size determination for studies of gene–environment interaction', *International Journal of Epidemiology*, 30: 1035–40.

Manson, J.E., Nathan, D.M., Krolewski, A.S., Stampfer, M.J., Willett, W.C. and Hennekens, C.H. (1992) 'A prospective study of exercise and incidence of diabetes among US male physicians', *JAMA*, 268: 63–7.

Marks, L. (1996) *Counting the Cost: The Real Impact of Non-insulin Dependent Diabetes*, London: King's Fund.

Marshall, J.A., Hamman, R.F. and Baxter, J. (1991) 'High-fat, low-carbohydrate diet and the etiology of non-insulin-dependent diabetes mellitus: the San Luis Valley diabetes study', *American Journal of Epidemology*, 134: 590–603.

McCance, D.R., Hanson, R.L., Charles, M.-A., Jacobsson, L.T.H., Pettit D.J. and Bennett, P.H. (1994) 'Comparison of tests for glycated haemoglobin and fasting and two hour plasma glucose concentrations as diagnostic methods for diabetes', *British Medical Journal*, 308: 1323–8.

Meirhaeghe, A., Helbecque, N., Cottel, D. and Amouyel, P. (1999) 'β_2-adrenoceptor gene polymorphism, body weight, and physical activity', *The Lancet*, 353: 896.

Meirhaeghe, A., Luan, J., Selberg-Franks, P., Hennings, S., Mitchell, J., Halsall, D., O'Rahilly, S. and Wareham, N.J. (2001) 'The effect of the Gly16Arg polymorphism of the β_2 adrenergic receptor gene on plasma free fatty acid levels is modulated by physical activity', *Journal of Clin Endocrinol Metab*, 86: 5881–7.

Morris, R.D., Rimm, D.L., Hartz, A.J., Kalkhoff, R.K. and Rimm, A.A. (1988) 'Obesity and heredity in the etiology of non-insulin dependent diabetes mellitus in 32,662 adult white women', *American Journal of Epidemiology*, 130: 112–21.

Murata, C., Suzuki, Y., Muramatsu, T., Taniyama, M., Atsumi, Y. and Matsuoka, K. (2000) 'Inactive aldehyde dehydrogenase 2 worsens glycemic control in patients with type 2 diabetes mellitus who drink low to moderate amounts of alcohol', *Alcohol Clinical and Experimental Research*, 24: 5s–11s.

Neel, J.V. (1962) 'Diabetes mellitus: a thrifty genotype rendered detrimental by "progress"?', *American Journal of Human Genetics*, 14: 353–62.

Nelson, R.G., Knowler, W.C., Pettit, D.J. and Bennett, P.H. (1995) 'Kidney diseases in diabetes', in National Diabetes Data Group (ed.) *Diabetes in America*, 2nd edn, Bethesda, MA: National Institutes of Health, pp. 349–85.

Pan, X.-R., Cao, H.-B., Li, G.-W., Hu, Y.-H., Jiang, X.-G., Wang, J.-X., Jiang, Y.-Y., Yang, W.-Y., Wang, J.-P., An, Z.-X., Zheng, H., Hu, Z.-X., Zhang, H., Juan-Lin, X., Bennett, P.H., Xiao, J.-Z. and Howard, B.V. (1997) 'Effect of dietary and exercise in preventing NIDDM in people with impaired glucose tolerance: the Da Qing IGT and diabetes study', *Diabetes Care*, 20: 537–44.

Phillips, A.N. and Davey Smith, G. (1993) 'The design of prospective epidemiological studies: more subjects of better measurements?', *Journal of Clinical Epidemiology*, 46: 1203–11.

Salas, J., Jansen, S., Lopez-Miranda, J., Ordovas, J.M., Castro, P., Marin, C., Ostos, M.A., Bravo, M.D., Jimenez-Pereperez, J., Blanco, A., Lopez-Segura, F. and Perez-Jimenez, F. (1998) 'The SstI polymorphism of the apolipoprotein C-III gene determines the insulin response to an oral-glucose-tolerance test after consumption of a diet rich in saturated fats', *American Journal of Clinical Nutrition*, 68: 396–401.

Sargeant, L.A., Wareham, N.J. and Khaw, K.-T. (2000) 'Family history of diabetes identifies a group at increased risk for the metabolic consequences of obesity and physical inactivity in EPIC-Norfolk: a population-based study', *International Journal of Obesity*, 24: 1333–9.

Smith, P.G. and Day, N.E. (1984) 'The design of case-control studies: the influence of confounding and interaction effects' *International Journal of Epidemiology*, 13: 356–65.

Stanhope, J.M. and Prior, I.A.M. (1980) 'The Tokelau Island Migrant Study: prevalence and incidence of diabetes mellitus', *New Zealand Medical Journal*, 92: 417–21.

Swinburn, B.A. (1996) 'The thrifty genotype hypothesis: how does it look after 30 years?', *Diabetic Medicine*, 13: 695–9.

Tattersall, R.B. and Pyke, D.A. (1972) 'Diabetes in twins', *Lancet*, ii: 1120–5.

The Lower Extremity Amputation Study Group (1995) 'Comparing the incidence of lower extremity amputations across the world: the Global Lower Extremity Amputation Study', *Diabetic Medicine*, 12: 14–18.

Tuomilehto, J., Lindström, J., Eriksson, J.G., Valle, T.T., Hämäläinen, H., Ilanne-Parikka, P., Keinänen-kiukaanniemi, S., Laakso, M., Louheranta, A., Rastas, M., Salminen, V. and Uusitupa, M. (2001) 'Prevention of type 2 diabetes mellitus by changes in lifestyle among subjects with impaired glucose tolerance', *New England Journal of Medicine*, 344: 1343–50.

Ukkola, O., Sun, G. and Bouchard, C. (2001) 'Insulin-like growth factor 2 (IGF2) and IGF-binding protein 1 (IGFBP1) gene variants are associated with overfeeding-induced metabolic changes', *Diabetologia*, 44: 2231–6.

Wareham, N.J., Byrne, C.D., Williams, R., Day, N.E. and Hales, C.N. (1999) 'Fasting proinsulin concentrations predict the development of type 2 diabetes', *Diabetes Care*, 22: 262–70.

Wareham, N.J., Wong, M.-Y. and Day, N.E. (2000) 'Glucose intolerance and physical inactivity: the relative importance of low habitual energy expenditure and cardio-respiratory fitness', *American Journal of Epidemiology*, 152: 132–9.

Wendorf, M. (1989) 'Diabetes, the ice free corridor, and the Paleoindian settlement of North America', *American Journal of Physical Anthropology*, 79: 503–20.

Williams, D.R.R., Wareham, N.J., Brown, D.C., Byrne, C.D., Cox, B.D., Day, N.E., Shackleton, J.R., Wang, T.W.M. and Hales, C.N. (1995) 'Glucose intolerance in the community; the Isle of Ely Diabetes Project', *Diabetic Medicine*, 12: 30–5.

Zimmet, P., King, H., Taylor, R., Raper, L.R., Balkau, B., Borger, J., Heriot, W. and Thoma, K. (1984) 'The high prevalence of diabetes mellitus, impaired glucose tolerance and diabetic retinopathy in Nauru – the 1982 survey', *Diabetes Res*, 1: 13–18.

9 Investigating the 'hidden' epidemic

Sexual behaviour and representations of HIV/AIDS amongst business people and medical personnel in five Central and Eastern European nations

Robin Goodwin

Introduction

As the HIV/AIDS epidemic sweeps the world, and the number of adults and children infected by the HIV virus tops 34 million (Joint United Nations Programme on HIV/AIDS, 2000a) the spread of the epidemic in Central and particularly Eastern Europe has received relatively little attention (Barnett *et al.*, 2000). Whilst the HIV/AIDS epidemic is a relatively recent phenomenon in Eastern Europe, only beginning in the early 1990s, WHO AIDS surveillance figures indicate a rapid growth in both HIV and AIDS in Eastern Europe over the past five years, with Central and Eastern Europe now showing the world's steepest HIV curve (European Centre for Epidemiological Monitoring of AIDS in Europe, 1999; Joint United Nations Programme on HIV/AIDS, 2000b). In the towns and cities surrounding Moscow, HIV infection increased five times in the first nine months of 1999 (Joint United Nations Programme on HIV/AIDS, 2000a) and new infections during the year 2000 were higher than in all previous years of the epidemic combined (Joint United Nations Programme on HIV/AIDS, 2000b). Meanwhile the geographical location of other, less-infected nations in the former Soviet Bloc on drug routes, and the high rates of population movement across this region, raise the prospect of a rapid increase in the spread of the epidemic (Dehne *et al.*, 1999).

A number of social and structural factors have been suggested to explain the rapid rise of HIV infection in these countries. Such explanations include the growth of temporary sexual partnerships as a means of economic survival or as a response to coping with a stressful environment (Kalichman, 1998) and the continuing acceptance of sexual violence in many of these societies, a factor contributing to high infection risk (Kalichman *et al.*, 2000). At the same time, a widespread belief that HIV is an 'outsiders' problem associated with the 'decadent West' has led to controversial legislation which reflects the political sensitivity of the epidemic. A new AIDS law introduced in Russia during the mid-1990s required the compulsory testing of foreigners visiting Russia for more than three months, reminiscent of calls for isolation common to epidemics across the ages (Sontag, 1989). In Georgia, an article in *Akhali Taoba* (3 December 1996) reporting a growing numbers of AIDS cases was contradicted a week later by the Department of Social and Economical Information which reported there were 'no' AIDS cases in the country (*Kavkasioni*, 11 December 1996).

As yet there has been little empirical research examining the individual values or the social representations that might underlie sexual behaviour in this region. Despite only limited evidence that AIDS-related knowledge and attitudes shape sexual practice, most of the existent research provides cognitive explanations which attempt to link individual knowledge and attitudes to sexual behaviour (Campbell, 2001). A particular problem with this knowledge-based paradigm is the assumption that individuals make considered, rational health choices, whereas sexual activity is often highly emotional and involves a range of social, often unconscious factors (Breakwell *et al.*, 1994). Furthermore, knowledge-based approaches explain little of the 'moral panic' that follows the spread of such a sexual epidemic. By permitting individuals to psychologically 'distance' themselves from high-risk groups, such individuals often ironically expose themselves to further risk (Lear, 1995).

In the project described in this chapter two theoretical frameworks are adopted to investigate HIV/AIDS in Central and Eastern Europe. The first approach focused on the role played by individual values in sexual behaviours in this region, exploring the relationship between these values and sexual behaviours (Goodwin *et al.*, in press). The second theoretical approach was drawn from the theory of social representations (Moscovici, 1984; Goodwin *et al.*, submitted). Social representations theory examines the way in which folk theories and everyday knowledge serve vital social functions in guiding and justifying actions, maintaining social identity and allowing for the communication between group members (Páez *et al.*, 1991). Here the particular interest was in the words and concepts that individuals most closely associate with HIV/AIDS, as well as participants' responses to a series of interview questions aimed at exploring their 'common sense' percep-tions of the epidemic in this region.

The data reported throughout this chapter was collected during 1998 and 1999 in Estonia, Georgia, Hungary, Poland and Russia. Data were collected in the capital cities of Hungary, Russia, Georgia and Estonia with additional data collected in the relatively high infection areas of Eastern Poland, in St. Petersburg, Russia and in the towns of Kutaisi and Batumi (Georgia) and Tartu (Estonia). The countries studied vary not only in the spread of the epidemic in each nation but in political structure, influence and nature of the religion(s) practised and level of economic investment and growth – all factors likely to have important implications for the spread of sexual infection (Borisenko *et al.*, 1999). Estonia and Georgia were both parts of the Former Soviet Union with low rates of HIV infection (less than 50 cases of AIDS had been reported in each country at the beginning of the study). In Hungary and Poland, rates of infection were also relatively low and stable, with little evidence of a marked increase in the prevalence of HIV/AIDS (Poland: 794 AIDS cases;

Hungary: 328 AIDS cases; European Centre for Epidemiological Monitoring of AIDS in Europe, 1999). In contrast, the Russian Federation had seen a marked escalation in the HIV epidemic, with HIV infection now increasing at one of the fastest rates in the world (Joint United Nations Programme on HIV/AIDS, 2000a).

The analyses were conducted with two different groups of participants: medics and business people. Business people were selected as a highly mobile group whose lifestyle and relatively high income permits them to engage in particular, higher risk activities (Barnett *et al.*, 2000). Such a group are particularly likely to visit sex workers (Wellings *et al.*, 1994), an important risk factor in this region for HIV infection (Towianska *et al.*, 1992). Given the inconsistent levels of screening and generally poor hospital conditions in these nations (Renton *et al.*, 1999), medics were also viewed as a relatively high-risk group. Despite widespread evidence of considerable 'misrepresentations' of HIV amongst medical professionals across a host of societies (Markova and Wilkie, 1987; Echebarria and Páez Rovira, 1989), the representations of medical staff are likely to be highly influential in these transient societies (Rivkin-Fish, 1999).

Study 1: values and sexual behaviour

During the last decade, Shalom Schwartz and his team have developed a quasi-circumplex model of ten values whose structure has been verified in more than 70 cultures, including the five included in the present study (Schwartz, 1992; Schwartz and Sagiv, 1995). The model can be organised along two dimensions: (i) Openness to Change (independent thought and action, reflected in concepts of Autonomy) versus Conservatism (Conformity, Tradition and Security) and (ii) Self-Transcendence (Universalism and Benevolence) versus Self-Enhancement (Achievement and Power). A final value type, Hedonism, is related to both Openness to Change and Self-Enhancement.

This study built on previous work on values by Schwartz and his team, as well as previous research on sensation seeking and sexual adventurism, to suggest a number of hypotheses. First, it was predicted that respondents who score highly on the Openness to Change dimension will stress a pleasurable, varied and daring life. It was predicted that those who score highly on the Openness to Change value dimension will have had more sexual encounters and will report more sexual disease than those emphasising the Conservatism dimension (Conformity, Tradition and Security). Second, it was expected that those scoring high on Hedonism will also seek pleasure and sensual gratification for themselves, and embrace risk taking. Just as sexual adventurists are likely to be at high risk of HIV infection, it might also be expected that those high on hedonism report more sexual disease and more multiple partnerships. Finally, it was predicted that the relationship between Self-Enhancement/Self-Transcendence and sexual behaviour is likely to be complex and gender specific. Sociobiological research has argued that a strong emphasis on social prestige and power amongst men can act as an important marker for female mate attraction, but that such a preference is less likely for males seeking female partners (Buss, 1989). Therefore a positive correlation between self-enhancement and number of sexual partners amongst male (but not female) respondents was anticipated.

Method

Altogether, 503 persons (51 per cent business people, 20 per cent nurses and 29 per cent doctors) participated in the first study of values and sexual behaviour. One hundred participants were recruited in Hungary, Poland and Russia; 101 in Estonia and 102 in Georgia. Participants were recruited primarily through medical schools or business training courses taught by the research team or their colleagues. Respondents were generally well educated, with 62 per cent having completed university, although only 38 per cent of nurses had continued beyond the 'special secondary' level of education. Polish respondents were overwhelmingly Catholic (93 per cent of the sample) and a further 53 per cent of Hungarian respondents were also Catholics. The Orthodox Church was strongly represented in Georgia (96 per cent of respondents were Georgian Orthodox) and Russia (75 per cent of participants were Russian Orthodox). Seventy-one percent of the Estonian respondents were not religious.

Values were assessed using a version of Schwartz's Portrait Values Scale (Schwartz *et al.*, 2001), which provided 29 descriptions of imaginary individuals and asked 'How much is the person in the description like you?' Scale items were divided into the ten value types, then formed into the dimensions Openness to Change versus Conservatism and Self-Transcendence versus Self-Enhancement. To indicate sexual behaviour, respondents reported whether they had ever had sexual intercourse, the number of sexual partners they had had over the past week/month/six months, and their frequency of condom use over the past six months (on a 4-point scale ranging from never to always). They also indicated whether they had ever had a sexually transmitted diseases STD (scored on a binary scale). Additional binary variables were also created for the presence (absence) of multiple partnerships over the previous six months (RISKY) and the presence (absence) of multiple partnerships accompanied by low (high) condom usage (VRISKY). To deal with possible cultural biases in question design and implementation procedures, items were extensively discussed in a group meeting in Estonia, reworded to maximise cultural sensitivity and then back-translated by bilingual translators with revised versions checked for accuracy. Scales were also piloted on a sub-sample of ten medical workers and ten entrepreneurs in each culture to remove any further ambiguities prior to the main data collection.

Results

Sexual behaviour across the samples

Ten percent of the sample reported that they were not sexually active over the past six months. Sixteen percent of the complete sample reported more than one sexual partner a month and seven percent more than one partner a week. The largest number of partners was reported by the Russian respondents (2.4 over six months: SD 2.99 range 0–20) and the lowest by the Polish participants (1.2 over six months: SD 1.38 range 0–10). Business people ($M = 1.8$) recorded more sexual partners than the other two occupational groups during this six-month period (M for doctors = 1.7, nurses = 1.1). Seventeen percent of the sample had had a previous STD, with a sexual disease reported more frequently by Russian, Georgian and Estonian respondents than by Hungarian or Polish respondents. Business people and doctors rather than nurses were more likely to report STDs.

Values and sexual behaviour

First, raw correlations were calculated between the ten individual value types and two dimensions, the number of partners over the previous six months and the binary variables described above (RISKY, VRISKY and STD). The findings indicate relatively small but consistent correlations between values and sexual behaviour. Self-Direction, Stimulation, Hedonism, Achievement and Power were all predictors of generally riskier sexual behaviour whilst Universalism, Benevolence, Conformity, Tradition and Security were indicators of safer sexual practices. Number of partners over the previous six months was positively correlated to the individual values of Power, Achievement, Hedonism and Stimulation and negatively correlated to values of Universalism, Benevolence, Conformity, Tradition and Security. Eight of these ten values (all but Conformity and Self-Direction) were significant indicators of the combined variable 'VRISKY' (i.e. multiple partnerships and low condom use). Those higher in Power, Achievement, Hedonism, Stimulation and Self-Direction were also significantly more likely to indicate a history of STD: those higher in Universalism, Conformity, Tradition and Security were less likely to report such disease.

Above it was suggested that those high on the dimension Openness to Change and high on Hedonism would have had more sexual encounters and a greater exposure to STD. A series of analyses of variance revealed that those who had multiple partners over the previous six months and only rarely used a condom were higher on Hedonism and Openness to Change. These 'high-risk' individuals were also higher on Self-Enhancement, but, contrary to our third expectation, this pattern was evident for both sexes. Sexual diseases were more likely to be reported by those scoring high on Hedonism, Openness to Change and Achievement and Power.

Discussion of study 1

The first study examined the relationship between ten value types, condom use, history of sexual disease and number of sexual partners in an area of rising HIV infection. The data demonstrated that values were moderate but significant predictors of reported sexual behaviours, with riskier sexual activity reported by those who were Open to Change, and high on Hedonism and Self-Enhancement.

Reported sexual activity varied by both occupation and nation. Most notably, sexual activity was highest amongst the Russian respondents. Whilst the high rates of multiple partnership were perhaps not surprising given the currently high rates of sexual disease in this country, the occurrence of high-risk behaviour amongst the highly qualified doctors in this nation did surprise us (more than 20 per cent of Russian doctors reported more than one sexual partner per week). In contrast, the Polish sample reported very few multiple partnerships and low rates of sexual disease. Poland can be characterised as a 'familistic' culture (Fukuyama, 1995) where family-centred social networks may serve to inhibit casual sexual activity. Religion may be an additional factor in sexual behaviour in this culture (Weinberg *et al.*, 2000); nearly all the Polish respondents were Catholics with the Catholic Church in Poland having recently played a highly visible – if somewhat controversial – role in the promotion of safer sexual practices in this nation (Danziger, 1996).

The first analysis was aimed at the 'individual level' of analysis and was less concerned with group representations of the HIV/AIDS epidemic in this region. In contrast, studies of HIV from a social representations perspective have been particularly concerned with the way in which different groups protect their different identities throughout – group-specific representations (Pàez *et al.*, 1991), with many early representations placing the epidemic within fringe, pariah groups whose indulgent, immoral behaviour was responsible for their infection (Echebarria and Páez Rovira, 1989; Sontag, 1989). This is likely to be particularly pertinent for the situation in Central / Eastern Europe, in which increases in HIV/AIDS infection have been accompanied by widespread discrimination against those infected (Danziger, 1994).

Study 2: representations of HIV/AIDS

The second study was concerned with the content of social representations of HIV/AIDS, tapped primarily through an analysis of the discourses that surround this epidemic. The study can be divided into two parts. First, respondents were asked a total of fifteen questions taken primarily from the work of Joffe (1996). Questions were grouped into three clusters (see Appendix): four questions concerned the origin and spread of HIV/AIDS, six concerned the nature of high-risk groups, and five concerned the government's role in caring for those infected by HIV. In a second task, respondents were instructed to write 'everything that comes into your mind when I say the word AIDS' (after Doise *et al.*, 1993). Respondents were asked to write as many responses as possible and to be 'unrestricted' in their responses. In total, 511 respondents took part in this second study of which 50 per cent were business people, 20 per cent nurses and 30 per cent doctors. Respondents for this second study were from a very similar background to those that took part in the values analysis described above, although only one third of the participants in the first study participated in the second phase of this research.

Study 2a: Interview analysis

Analysis was through a multi-stage procedure. First, interviews were tape-recorded and transcribed by the researchers in each country. Second, researchers in each country 'immersed' themselves in this data to allow us to identify key categories of response for each question. Third, lists of key responses were brought together in a week-long group meeting in Kutaisi, Georgia, attended by the full research team, allowing us to construct a coding scheme for each question. Coding was guided by a desire to capture both a common 'core' of frequent responses whilst not excluding less frequent responses of additional theoretical concern (for example, where certain sub-groups were seen as high risk by only sections of the sample). Below, a brief summary of responses grouped for the three sets of questions asked, is presented. A full list of codings for each question is available from the author.

The origin and spread of HIV/AIDS

Respondents were generally very concerned about the spread of HIV/AIDS with the great majority (85 per cent) of interviewees viewing HIV/AIDS as a serious, global problem. Only in one country, Russia, did respondents see the epidemic as having 'little to do with them' (13 Russian business people and 11 medics made this claim) although in Hungary, Russia and Poland more than 20 per cent of respondents thought the magnitude of the problem had been 'exaggerated'. Considerably more Russian respondents (30 in total) saw the epidemic as a 'solvable problem'.

Respondents learned about HIV/AIDS primarily through the media (71 per cent of respondents claimed this), with few respondents (4 per cent) reporting that they first learned of the epidemic through their colleagues or friends. Georgian medics were the least likely to report they had heard of the epidemic first in their work environment. Africa was viewed as the origin of HIV/AIDS by the majority (68 per cent) of respondents. However, a sizeable proportion of medical respondents in Poland (16 per cent) and Russia (20 per cent) also saw the virus as emanating from 'the West'. In addition, more than 10 per cent of respondents in Georgia, Estonia and Russia were willing to cite 'conspiracy theories' concerning the origin of the virus (e.g. HIV was developed for biological warfare purposes). AIDS was seen as spreading primarily through sex and blood with little mention of the risks of mother–child transmission outside of the Hungarian medical sample. The risk of contamination through 'syringes' was mentioned by more than half the Polish respondents and 40 per cent of Russian respondents but was not widely noted by other respondents.

'High-risk' groups

The data provided only partial evidence of a clear 'out-grouping' in response to the HIV epidemic. The majority of the sample were unwilling to directly place blame on the individual for contracting AIDS (only 98 respondents directly answered 'yes' to a question asking 'Is contracting AIDS the person's own fault?'; Q7) although a further 144 were prepared to place conditional blame dependent on circumstances. Most respondents did not feel that people were particularly safe because of a high level of morality in their own country (Q10) – this was particularly notable in Russia, where 42 business people and 39 respondents categorically denied this.

Despite this, particular groups were identified as being particularly risky in each country, and the extent to which 'everyone' was seen as at risk ranged from 11 per cent of respondents in Georgia to 57 per cent in Hungary (Q5). Prostitutes in Estonia and Georgia, 'young' people in Poland and the 'socially maladjusted' in Poland and Georgia were seen as relatively high risk. In response to the question 'What kind of person carries a condom?' (Q9) the 'educated' or 'cautious' majority were identified in each country, although notably in Poland approximately 10 per cent of respondents identified 'modern young women' who carried condoms as members of a relatively 'immoral' group of individuals at risk from infection.

The role of wider society

The vast majority of respondents in each sample (83 per cent of respondents overall) saw government as having a prominent role in dealing with the HIV epidemic (Q11). However, 16 per cent of Russian business people and 14 per cent of Russian medical workers saw the responsibility as laying elsewhere and some 12 per cent of Hungarians and 8 per cent of Georgian business people were keen to see a shared responsibility between the government, the individual and other agencies. A small number of respondents in Russia (13), Hungary (9) and Estonia (6) saw AIDS patients as having some responsibility to pay for their own treatments (Q12) and a further seven respondents in Poland gave a 'qualified' answer that payment should depend on the circumstances in which the disease was contracted. Where estimates were made of the number of those infected (Q13), Hungarian respondents were the most likely to exaggerate the numbers of those infected, with 24 per cent of business people and 18 per cent of medics giving figures that were higher than UN AIDS statistics for the time of data collection.

Finally, uncertainty about the respondent's own HIV status was lowest in Russia (Q14), reflecting the high levels of testing evident in this country. In Georgia and Poland, uncertainty about where to obtain an HIV test was most evident amongst the business community, with some 40 per cent of those questioned not knowing where they might be tested (Q15).

Study 2b: Free associations

Respondents also completed free associations to the words HIV/AIDS, with the 511 respondents producing a total of 1480 responses. Two judges examined a full and translated list of these associations and worked independently to aggregate semantically similar words identifying the thirteen most frequently listed words from across the sample. Table 9.1 provides percentages of respondents citing these thirteen words by country and occupational group.

The most frequently cited words across countries were disease and death, reflecting the generally pessimistic view of HIV and its outcomes reflected in the interview data. Drugs and blood were also frequently cited, reflecting widespread perceptions of the means of HIV transmission. Drugs were cited by Estonian participants in particular (representing 15 per cent of all Estonian responses to

Table 9.1 Free associations with AIDS/HIV: percentages of respondents citing these words rank-ordered by frequency

	Hungary		Georgia		Poland		Estonia		Russia		Raw total of responses given
	Business people	Medics	Business people	Medics	Business people	Medics	Business people	Medics	Business people	Medics	
Disease	62	44	27	23	64	48	74	68	70	59	270
Death	44	48	25	40	32	26	46	46	32	5	172
Drugs	22	24	20	34	42	30	66	62	12	8	160
Blood	32	20	14	19	61	52	64	58	0	0	160
Homosexual	42	40	14	21	32	40	48	44	10	4	148
Condoms	20	30	21	27	20	12	48	46	10	6	120
Sexual activity	20	30	0	25	46	30	42	42	0	0	117
Fear	30	22	37	42	0	0	0	0	58	23	106
Africa	18	10	0	0	32	40	20	12	0	0	66
Prostitution	14	14	0	0	40	30	20	0	10	4	66
Casual sex	10	22	0	0	36	14	0	0	0	0	41
Misfortune /intolerance	0	0	31	14	10	12	0	0	0	0	33
Hopelessness	0	0	0	0	0	0	10	0	14	19	21
Total no. of responses	158	152	95	123	208	167	219	189	108	61	1480

the free association task), whilst blood was not mentioned directly at all by the Russian respondents in their free associations. Homosexuality was more frequently mentioned by the Hungarian respondents (13 per cent of all Hungarian responses), whilst sexual activity in general was cited most by the Poles and Estonians (10 per cent of all responses for both national groups). Africa was more frequently cited by the Polish respondents (10 per cent of all responses), as was casual sex and prostitution (7 per cent and 9 per cent of total Polish responses). In Georgia, there was evidence of the most compassionate attitude towards AIDS sufferers, with a critique of the intolerance with which the infected were treated being cited by almost a third of the entrepreneurs and 14 per cent of the medics. Finally, the 'hopelessness' of the situation in Russia, the country with the most severe risk of the spread of infection at the time of questioning, was reflected in the high proportion of this response by the Russian respondents (10 per cent of the total of all Russian responses).

Discussion of study 2

The respondents demonstrated both similarities and differences in their representations of HIV/AIDS across the five cultures studied. From a structuralist perspective to social representations (Flament, 1994) a relatively stable and consensual key 'nucleus' around which the other representations were organised can be identified, and in our study it was clear that HIV was associated with a core set of very negative outcomes and images (death and disease were primary associations in all five countries, representing more than 30 per cent of all the free associations made). At the same time, however, it was clear that beliefs about the origins and spread of HIV, prevailing conceptualisations of moral responsibility and blame, and attitudes towards the role of the government were unevenly represented across the sample.

The conspiracy theories repeated by respondents were a prominent feature of the early Soviet reporting of HIV/AIDS (Headley, 1998; Sontag, 1989). In this context, therefore, it was perhaps unsurprising to see them still playing an important role in the representations of HIV in the three former Soviet nations (Georgia, Russia and Estonia). The data, however, provided only partial evidence of an 'out-grouping' in response to the HIV epidemic, although particular groups were more likely to be identified as being particularly 'risky' in each country. In particular, both the interview and free associations findings suggest conservative representations in Poland, and, to a lesser extent, Georgia, where images of the 'promiscuous' young condom carrier were accompanied with the strongest associations between casual sexual activity and HIV/AIDS (see Páez *et al.*, 1991 for examples of conservative representations of HIV in other cultures). At the same time it was in these same nations – Poland and Georgia – that there was evidence of the most compassionate attitude towards AIDS sufferers, demonstrating a more compassionate aspect to this conservative and moralistic tone (Páez *et al.*, 1991).

One intriguing group of respondents were the relatively high number of Russians who believed that the 'problem of HIV' could be solved or was exaggerated. This was accompanied by the third of Russian respondents who emphasised the 'hopelessness' of the situation in the free responses, reflecting an apocalyptic sense of doom frequently evoked in the early days of the Western epidemic and anchored in a wide range of fears about the future (Sontag, 1989). This bipolarity in perspectives – between relatively extreme poles of optimism and fatalistic 'AIDS-phobia' (*spidofobiia*) – was evident amongst both the medical and business sub-samples in this country. This ambiguity offers an intriguing topic for further study in this, the country currently most affected by this epidemic.

General discussion

This chapter describes part of a multi-stage project conducted to examine the representations of HIV/AIDS in five Central and Eastern European countries. Participants completed questionnaires measuring individual values and sexual behaviour and responded to interviews and free associations examining representations of HIV/AIDS in this region. The results indicated the importance of considering individual values when devising strategies aimed at reducing high-risk behaviours, as well as the significant cultural differences in representations of HIV and those seen as at-risk from infection. Notably, although respondents in the study were generally unwilling to condemn particular groups of respondents as being 'responsible' for the HIV epidemic in their country, specific groups were identified as being high risk in each country, with the 'socially maladjusted' being a particular target amongst Georgian business people and the 'asocial young' a particular focus of the Polish respondents. Whilst the origin of these images cannot be certain, the strong Catholic orientation amongst the Polish respondents seems consistent with the condemnation of condom carriers found in this culture.

What are the wider implications of our findings for the inhabitants of these societies? HIV is a lentivirus with a very long epidemic curve (Barnett, 2001). As a result it has long-term, often hidden impacts on the historical and development trajectories for a society (Barnett, 2001). High rates of morbidity and mortality in these poor and very unequal societies are likely to have profound impacts on a range of facets of social and economic life, with increased pressures on already stressed health services and increases in dependency ratios in households and numbers of orphans (Barnett *et al.*, 2000). In the country in the sample where the epidemic is now most widespread (Russia), a sizeable proportion of respondents saw the epidemic as having 'nothing to do with them', or, alternatively, seemed immobilised by a general sense of fatalism likely only to hinder safer sex behaviour. In the questionnaire study, high rates of sexual partnerships amongst the Russian respondents were demonstrated. Russia might be characterised as slipping into a 'risk society' (Barnett *et al.*, 2000), where the environment conspires to make activities risky and where sexual inhibitions, relaxed following the social transitions in this society, have accompanied the increasingly large number of Russians who have turned to sex for financial reasons (Headley, 1998).

The number of HIV/AIDS cases in Central and Eastern Europe is still not large compared with many other afflicted regions of the world. As a consequence, it should still be possible to limit the size of the epidemic in this region (Kalichman *et al.*, 2000). The findings emphasise the need to attract and hold the attention of target audiences by developing dynamic programmes targeted towards both the 'psychological profile' of these audiences and the wider representations of the epidemic held in these communities. Sub-cultures which stress values such as power and stimulation need to be targeted with attractive interventions which allow for the values of these groups, and the reaction of such sub-cultures to more conservative 'traditional' health messages needs to be anticipated. Differing representations imply new promotional campaigns that embody existing religious and social values (Páez *et al.*, 1991), as well as the wider socio-economic demands and ideological freedoms operating in post-Communist Europe. Finally, media interventions need to be constructed within the context of enduring economic realities, where the adoption of safer sexual behaviour may threaten not only romantic ties but also economic survival (Hobfoll, 1998). Any successful assault on this epidemic will therefore require both recognition of the role played by these values and their inter-relationship with the ecological realities of this troubled region.

Appendix: Interview questions

Group a. Origin and spread of HIV

1 What do you think of the AIDS problem generally?
2 How did you first hear about AIDS?
3 Where do you think AIDS originated?
4 How does AIDS spread amongst people?

Group b. 'Risk' groups

5 What type of person gets AIDS?
6 What types of person would you not have sex with?
7 Is contracting AIDS the person's own fault?
8 Do you know anyone personally who has AIDS or is infected?
9 What kind of person carries a condom?
10 Do you think your nation is safer from AIDS because of high sexual morality?

Group c. Society and government

11 Do you think it is the government's role to do something about AIDS?
12 Should the money for AIDS treatment come from taxpayers money? If not, from where?
13 How many people are infected with the AIDS virus in your country?
14 How do you know you are not infected with HIV?
15 Do you know where to get an HIV test?

Acknowledgements

This chapter describes part of a research project funded by the Soros Foundation, Prague. The research co-ordinators in each country were Anu Realo (Tartu University, Estonia); Lan-Anh Nguyen Luu (Eotvos Lorand University, Budapest, Hungary); George Nizharadze (Georgian Academy of Sciences, Georgia); Alexandra Kozlova (St Petersburg State University, Russia) and Anna Kwiatkowska (Warsaw School of Advanced Social Psychology, Poland).

References

Barnett, T. (2001) 'Divergent disciplinary and institutional approaches to the HIV/AIDS epidemic', paper presented at the Biosocial Society Symposium 'Learning from HIV/ AIDS: Transdisciplinary perspective', Institute of Education, London.

Barnett, T., Whiteside, A., Khodakevich, L., Kruglov, Y. and Steshenko, V. (2000) 'The HIV/AIDS epidemic in Ukraine: its potential social and economic impact', *Social Science and Medicine*, 51: 1387–403.

Borisenko, K.K., Tichonova, L.I. and Renton, A.M. (1999) 'Syphilis and other sexually transmitted infections in the Russian Federation', *International Journal of STD and AIDS*, 10: 665–8.

Breakwell, G.M., Millward, L.J. and Fife-Schaw, C. (1994) 'Commitment to 'safer' sex as a predictor of condom use among 16–20 year olds', *Journal of Applied Social Psychology*, 24: 189–217.

Buss, D.M. (1989) 'Sex differences in human mate preferences: evolutionary hypotheses tested in 37 cultures', *Behavioral and Brain Sciences*, 12: 1–14.

Campbell, C. (2001) 'More questions than answers? The impact of the HIV/AIDS epidemic on our understanding of social and community structures', paper presented at the Biosocial Society Symposium 'Learning from HIV/AIDS: Transdisciplinary perspective', Institute of Education, London.

Danziger, R. (1994) 'Discrimination against people with HIV and AIDS in Poland', *British Medical Journal*, 308: 1145–7.

Danziger, R. (1996) 'Compulsory testing for HIV in Hungary', *Social Science and Medicine*, 43: 1199–204.

Dehne, K.L., Khodakevich, L., Hamers, F.F. and Schwartlander, B. (1999) 'The HIV/ AIDS epidemic in Eastern Europe: recent patterns and trends and their implication for policy-making', *AIDS*, 13: 741–9.

Doise, W., Clemence, A. and Lorenzi-Cioldi, F. (1993) *The Quantitative Analysis of Social Representations*, London: Harvester Wheatsheaf.

Echebarria, A.E. and Páez Rovira, D. (1989) 'Social representations and memory: the case of AIDS', *European Journal of Social Psychology*, 19: 543–51.

European Centre for Epidemiological Monitoring of AIDS in Europe (1999) *HIV/AIDS Surveillance in Europe*, Quarterly Report, http://www.ceses.org/aids.htm

Flament, C. (1994) 'Consensus, salience and necessity in social representations: technical note', *Papers on Social Representations*, 3: 97–105.

Fukuyama, F. (1995) *Trust*, London: Penguin.

Goodwin, R., Realo, A., Kwiatkowska, A., Kozlova, A., Nguyen Luu, L.A. and Nizharadze, G. (2002) 'Values and sexual behaviour in Central and Eastern Europe', *Journal of Health Psychology*, 7(1): 45–56.

Goodwin, R., Kozlova, A., Kwiatkowska, A., Nguyen Luu, L.A., Realo, A., Külvet. A. and Rämmer, A. (submitted) *Social representations of HIV/AIDS in Central and Eastern Europe*.

Headley, D. (1998) *HIV/AIDS in Russia*, London: Charities Aid Foundation, Russian Office.

Hobfoll, S.E. (1998) 'Ecology, community and AIDS', *American Journal of Community Psychology*, 26: 133–44.

Joffe, H. (1996) 'AIDS research and prevention: a social representation approach', *British Journal of Medical Psychology*, 69: 169–90.

Joint United Nations Programme on HIV/AIDS (2000a) *Report on the global HIV/AIDS epidemic*, www.unaids.org/epidemic_update/report/index.html

Joint United Nations Programme on HIV/AIDS (2000b) *Aids epidemic explodes in Eastern Europe*, www.unaids.org/wac/2000/wad00/files/ruepr.html

Kalichman, S.C. (1998) *Preventing AIDS: A Sourcebook of Behavioral Interventions*, Mahwah, NJ: Lawrence Erlbaum.

Kalichman, S.C., Kelly, J.A., Sikkema, K.J., Koslov, A.P., Shaboltas, A. and Granskaya, J. (2000) 'The emerging AIDS crisis in Russia: review of enabling factors and prevention needs', *International Journal of STD and AIDS*, 11: 71–5.

Lear, D. (1995) 'Sexual communication in the age of AIDS: the construction of risk and trust among young adults', *Social Science and Medicine*, 41: 1311–23.

Markova, I. and Wilkie, P. (1987) 'Representations, concepts and social change: the phenomenon of AIDS', *Journal for the Theory of Social Behaviour*, 17: 389–409.

Moscovici, S. (1984) 'The phenomenon of social representations', in R.M. Farr and S. Moscovici (eds) *Social Representations*, Cambridge: Cambridge University Press.

Páez, D., Echebarria, A., Valencia, J., Romo, I., San Juan, C. and Vergara, A. (1991) 'AIDS social representations: contents and processes', *Journal of Community and Applied Social Psychology*, 1: 89–104.

Renton, A.M., Borisenko, K.K., Tichonova, L.I. and Akovian, V.A. (1999) 'The control and management of sexually transmitted diseases: a comparison of the United Kingdom and the Russian Federation', *International Journal of STD and AIDS*, 10: 659–64.

Rivkin-Fish, M. (1999) 'Sexuality education in Russia: defining pleasure and danger for a fledgling democratic society', *Social Sciences and Medicine*, 49: 801–14.

Schwartz, S.H. (1992) 'Universals in the content and structure of values: theoretical advances and empirical tests in 20 countries', in M.P. Zanna (ed.) *Advances in Experimental Social Psychology*, vol 25, pp. 1–65, New York: Academic Press.

Schwartz, S.H. and Sagiv, L. (1995) 'Identifying culture specifics in the content and structure of values', *Journal of Cross-Cultural Psychology*, 26: 92–116.

Schwartz, S.H., Melech, G., Lehmann, A., Burgess, S. and Harris, M. (2001) 'Extending the cross-cultural validity of the theory of basic human values with a different method of measurement', *Journal of Cross-Cultural Psychology*, 32: 519–42.

Sontag, S. (1989) *AIDS and its Metaphors*, New York: Doubleday.

Towianska, A., Rozlucka, E. and Dabrowski, J. (1992) 'Prevalence of HIV anti-bodies in maritime workers and in other selected population groups in Poland', *Bulletin of the Institute of Maritime and Topical Medicine in Gdynia*, 43: 19–24.

Weinberg, M.S., Lottes, I. and Shaver, F.M. (2000) 'Sociocultural correlates of permissive sexual attitudes: a test of Reiss' hypotheses about Sweden and the United States', *Journal of Sex Research*, 37: 44–52.

Wellings, K., Fields, J., Johnson, A.M. and Wadsworth, J. (1994) *Sexual Behaviour in Britain*, London: Penguin.

10 The evolution of disease and the devolution of health care for American Indians

Stephen J. Kunitz

Introduction

It has been recognized for several decades that there has been a dramatic transformation in the epidemiologic regimes of many populations around the globe, with a decline of infectious diseases, an increase in life expectancy, and an increase, both absolute and relative, of chronic degenerative and man-made diseases (Omran 1971). Among the most dramatically rapid of these transformations has been the one experienced by indigenous peoples of the advanced industrial nations, including Australia, Canada, New Zealand, and the United States (Kunitz 1994).

The topic of this chapter is the transformation that has occurred among Indians of the United States. The decline of infectious diseases in this population owes a good deal to the availability of free, high quality public and personal health services since the 1950s. For instance, immunization rates of Indian children aged 0–27 months are higher than the rates for non-Indian children in the United States (IHS 1998–99). However, the increasing relative and absolute importance of some non-infectious conditions reflects in part a failure of that same health care system to address emerging problems in a timely fashion. This is not a unique situation, for in general health care systems have dealt more effectively with the infectious diseases of public health concern than they have with the non-infectious diseases.

Part of the reason that Indian health programmes have not dealt effectively with these problems is budgetary; health care costs have been increasing more rapidly than the money available for prevention and treatment. But another part of the reason has to do with the fact that at the same time as population-based primary prevention has received less attention than it deserves, genetic theories are becoming the dominant mode of explanation of non-infectious diseases. These various factors reinforce one another in ways suggested below.

Patterns of Indian health

The major causes of morbidity and mortality among American Indians have changed dramatically since World War II. Table 10.1 displays the ten leading causes of death in order of importance in 1951–53 and 1994–96, as well as life expectancy for Indian and non-Indian men and women in each period. It is evident

Table 10.1 Ten leading causes of death of American Indians in order of importance and life expectancy, 1950s and 1990s

Cause of death	
1951–1953	*1994–1996*
Heart disease	Diseases of the heart
Accidents	Cancer
Influenza and pneumonia	Accidents
TBC	Diabetes mellitis
Certain diseases of early infancy	Chronic liver disease and cirrhosis
Cancer	Cerebrovascular disease
Gastritis etc	Pneumonia and influenza
Vascular lesions affecting the CNS	Suicide
Congenital malformations	COPD
Homicide and legal execution	Homicide and legal execution

	Life expectancy at birth			
	1951–1953		*1994–1996*	
	Indians	*White*	*Indians*	*All races*
Men	58	67	67.6	72.5
Women	62	72	74.7	78.9

Sources: *Health Service for American Indians.* Division of Indian Health, U.S. Department of Health, Education and Welfare, Public Health Service Publication No. 531. Washington, D.C.: U.S. Government Printing Office, 1957; *Trends in Indian Health 1998-99.* Indian Health Service, Office of Public Health, Program Statistics Team, U.S. Department of Health and Human Services, Rockville. MD, 2000.

that even in the 1950s non-infectious diseases had become significant, but they had become even more so forty years later.

These data are somewhat misleading, however, for they mask enormous variation within the Indian population. Variations are not new, of course. They result from differences in the cultures and social organization of native populations, differences among colonists, and differences in the history of contact and economic growth. There were very substantial differences across the Americas with regard to the demographic and epidemiologic response to European contact beginning in the 16th century (Verrano and Ubelaker 1992). In the late 19th century there were very great variations in homicide rates among North American Indians (Levy and Kunitz 1971). During the early 20th century demographic recovery differed among Indian populations (Shoemaker 1999). In the 1950s the Indian Health Service documented very large variations in age-adjusted death rates among Indians. In the 1970s and 1980s there was very considerable variation in income and poverty among Indian reservations (Trosper 1996), and variation in measures of health persists right into the present, as the data in Table 10.2 indicate. Life expectancy among both Indian women and men differs by as much as ten years in different parts of the country.

Analyses of the data in Table 10.2 indicate that the socio-economic indicators

Table 10.2 Indian Health Service regional data, 1990s

Area	Individuals aged 35–44 using tobacco (%)	Lung cancer death rate*	Cancer death rate*	Heart disease death rate*	Diabetes death rate*	Alcoholism death rate*	Individuals aged 25 or over graduated from high school (%)	Individuals aged 25 or over graduated from college (%)	Males aged 16 or over unemployed (%)	Females aged 16 or over unemployed (%)	Median household income ($)	Above poverty line (%)	Female life expectancy	Male life expectancy
Aberdeen	55.0	50.6	172.9	229.7	68.7	108.7	64.4	7.8	26.5	19.4	12,310	49.6	70.0	60.6
Alaska	51.0	56.3	160.9	151.6	10.9	72.1	63.1	4.1	27.3	16.1	24,216	24.0	73.0	65.6
Albuquerque	22.0	8.4	90.0	85.1	74.6	70.7	61.2	7.0	20.0	15.8	15,791	42.1	75.8	69.9
Bemidji	65.0	78.4	216.2	287.0	78.3	39.2	67.4	7.2	18.5	14.3	19,317	33.0	67.8	62.4
Billings	40.0	69.7	173.3	206.4	64.5	60.6	67.6	7.6	29.8	21.0	14,249	44.6	70.8	63.6
California	38.0	26.5	78.0	129.3	28.0	27.0	71.4	11.1	11.6	10.4	28,029	24.0	80.2	72.5
Nashville	52.0	25.5	102.2	190.4	59.9	30.8	61.2	10.5	10.7	10.6	21,265	24.7	76.0	69.4
Navajo	16.0	6.8	85.5	105.7	42.5	50.1	54.8	5.2	23.5	18.6	13,984	46.8	76.7	68.3
Oklahoma	50.0	31.8	113.3	163.6	38.1	21.7	69.2	11.4	12.0	11.3	19,750	27.0	77.5	70.8
Phoenix	30.0	13.7	88.2	145.9	70.0	72.1	59.4	6.3	21.0	17.4	16,392	41.8	72.3	65.8
Portland	48.0	42.1	126.4	140.9	35.7	56.0	71.5	8.6	16.1	13.3	21,123	29.2	73.4	68.6
Tucson	?	11.2	99.4	137.5	79.7	70.0	52.1	4.6	25.2	20.2	13,342	24.0	72.0	62.2

Source: *Regional Differences in Indian Health, 1998–9*, Indian Health Service, Public Health Service, Department of Health and Human Services, Rockville, MD, 1999.

Note: * Death rates expressed as deaths per 1000,000.

are all associated, as one would expect. Unemployment is inversely correlated with education ($r = -0.79$ and -0.74) and median household income ($r = -0.61$). Of all the socio-economic indicators, unemployment is most strongly correlated with life expectancy, especially for men ($r = -0.74$). Interestingly, unemployment is only strongly correlated with the alcoholism death rate ($r = 0.75$). Tobacco use among 35–44 year olds is strongly correlated with deaths from lung cancer ($r = 0.77$) and heart disease ($r = 0.81$).

It is not my intention to account for these variations here, but only to make the following two points. First, the variations are substantial and are only partly associated with socio-economic indicators. Several very important causes of morbidity and mortality, as well as smoking, which is an important risk factor for several causes of morbidity and mortality, are not correlated – at least at the ecological level – with socio-economic conditions. Second, these very large differences in morbidity and mortality suggest the appropriateness of different types of responses on the part of providers attentive to the special needs of the particular populations they serve.

The impact of health care

As stated above, the decline of mortality caused by infectious diseases was to a considerable degree the result of the availability of a well-organized health care system which provided both public and personal preventive and therapeutic interventions (Kunitz 1983). It is also true that the health care system has not dealt successfully with the newly important diseases. This is not simply because such conditions are not amenable to interventions but because the necessary interventions have not been adapted and developed for the wide variety of American Indians.

A significant body of work over the past 20–30 years has been devoted to the assessment of the medical contribution to the improvement of the health of contemporary populations (Rutstein *et al.* 1976). Most of the studies in this tradition classify causes of death as either amenable or not amenable to medical care.

> Here medical care is defined in its broadest sense, that is prevention, cure and care, including the application of all relevant medical knowledge, the services of all medical and allied personnel, the resources of governmental, voluntary, and social agencies, and the co-operation of the individual himself. An excessive number of such unnecessary events serves as a warning signal of possible shortcomings in the health care system, and should be investigated further.
>
> (Holland 1993)

Avoidable deaths may thus arise for a variety of reasons, including inadequate funding, inaccessible services and/or populations, incompetent staff, or non-compliant patients. Though all of these factors may be contributory, the fact that some populations have higher rates than others is an indication that adequate

Table 10.3 Causes of death amenable to medical intervention: ratio of age adjusted death rates, Native Americans to non-indigenous Americans

	Indigenous Americans / Non-indigenous Americans	
	All Races	*White*
All ages		
Major cardiovascular		
Diseases of the heart	1.1	1.2
Cerebrovascular	1.1	1.2
Atherosclerosis	1.1	1.2
Hypertension	0.9	1.2
Motor vehicle accidents	3.3	3.3
Chronic liver disease and cirrhosis	4.9	5.0
Diabetes	3.5	4.0
Infant mortality	1.2	1.5
Maternal mortality	0.9	1.5
Aged over 55		
Breast cancer	0.6	0.6
Cervical cancer	1.8	1.9
Trachea, bronchus and lung cancer	0.8	0.8

Source: *Trends in Indian Health 1998–99.* Indian Health Service, Office of Public Health, Program Statistics Team, U.S. Department of Health and Human Services, Rockville. MD, 2000.

health services responsive to the unique needs of particular populations may not be available.

Table 10.3 displays data on deaths due to those leading causes generally thought to be amenable to medical intervention in the sense described above. Diabetes mellitus has been added to the list because preventive and therapeutic interventions are known, at least in broad outline (Broussard *et al.* 1995). The figures are the ratios of the age-adjusted death rates of American Indians and Alaska Natives to those of all non-Indians in the United States and to only white Americans.

Considering all native Americans together, the death rates from the major cardiovascular diseases are not very much higher than they are for other Americans. On the other hand, deaths from motor vehicle accidents, chronic liver disease and cirrhosis, and diabetes are all very much higher than in the non-Indian population. Cervical cancer and infant and maternal mortality also tend to be higher among Indians than non-Indians. Clearly, not all the responsibility for the differences, nor all the credit for the near equality of rates of death for the major cardiovascular diseases, can be laid at the door of the health care system. It seems, however, that some of the great discrepancies in several of these causes are at least partially the responsibility of the health care system, for two reasons.

First, there has been a failure to effectively apply knowledge accumulated over many years regarding the risk factors for some important conditions. For example, millions of dollars have been spent over the past thirty years on the study of non-insulin dependent diabetes mellitus (type 2 diabetes) among Indians in both the

United States and Canada. Until relatively recently, however, there have been few studies of preventive interventions, and the rate of type 2 diabetes has continued to increase. To take another example, despite very critical assessments of programmes for the prevention and treatment of alcoholism, existing programmes still are often highly disorganized, with high rates of staff turnover and low rates of success (May 1986; Kunitz and Levy 2000), and death rates from alcohol-related conditions have remained high.

Second, as already noted, health care for American Indians has been woefully under funded. Per capita expenditures on health services for Indians have been between 39 and 60 per cent of what is spent on the average American citizen, depending upon the source of data, and expenditures are not increasing rapidly enough to catch up (Kunitz 1996; Noren *et al.* 1998). One result is that money and personnel are not available in sufficient amounts to encourage different Indian tribes (with or without outside researchers and consultants) to engage in long-term prevention research projects in their own populations.

The provision of health services to American Indians and Alaska Natives

Many American Indians, especially those west of the Mississippi, have received free health care from the federal government since the 19th century. From the late 19th to the mid-20th century, care was provided by the Bureau of Indian Affairs, part of the Department of the Interior. Since the mid-1950s, care has been provided by the Indian Health Service, part of the U.S. Public Health Service that is located within the Department of Health and Human Services. Reasons for the transfer have been described in detail elsewhere (Kunitz 1996). It suffices to say here that federal Indian policy during the 1950s aimed at termination of the special status of Indians and their assimilation into the larger population. It was believed this could not be achieved as long as Indian health was so much worse than that of the rest of the population. In an effort to both weaken the Bureau of Indian Affairs and improve health care for, and the health status of, Indians transfer was mandated to what was then known as the Division of Indian Health. The clearly stated assumption was that once health had been improved sufficiently, the federal government would withdraw from the provision of free services and Indians would be dealt with in just the same way as every other part of the American population. That is to say, health care for American Indians is not an entitlement like Medicare and Social Security (Cunningham 1996).

Beginning in the early 1970s under the Nixon Administration, significant changes occurred in Indian policy. The new policy meant two things: (1) repudiation of termination and recognition of the rights of Indian tribes to self-determination, and (2) increased funding for Indian programmes (Kunitz 1996). One of President Nixon's staff, John Erlichmann, explained the President's interest in Indians as follows:

Nixon did have a personal interest in Indian issues. There were three reasons. First, he was a "strict constructionist" who believed that treaties were meant to be observed. Second, he believed that because they were relatively few in number, Indians were a manageable minority and that their problems *could* be addressed by the government. Finally, he was favourably disposed towards Indians because of his high regard for his football coach at Whittier, "Chief" Newman.

(Bergman *et al.* 1999)

The description of Indians as a "manageable minority", implies more than that their small numbers meant that the federal government could successfully address their problems. It also implies that they were not numerous enough to be seriously troublesome like African-Americans, and that devoting special attention to them would result in substantial gains politically for the administration, and economically and in terms of improved health for Indians, at relatively little cost.

Whatever the reasons, increased federal spending on health and other programmes contributed significantly to an increase in per capita income and a decline in the proportion of reservation populations living in poverty during the 1970s (Trosper 1996). Self-determination has manifested itself in increasing transfer to Indian tribal governments of responsibility for the management of many programmes previously run by the federal government, including of course health services (Adams 2000).

In addition to its consistency with the policy of self-determination, devolution of control from the federal to tribal governments is also consistent with the general policy of "outsourcing" in both the public and the private sectors of the economy. In the public sector, this is generally called "privatization". It is occurring widely throughout the United States, including in the public health arena where local health departments are increasingly relying on private contractors to provide services, especially clinical services, once provided by health departments themselves (Keane *et al.* 2001a, 2001b).

The consequences of outsourcing in general, and of devolution of health care for Indians in particular, appear to be mixed (Noren *et al.* 1998). When funding is sufficient, tribally managed programmes may be very innovative and responsive to the special needs of their particular populations. In addition to the arguments in favour of tribal sovereignty, this is of great potential value given the enormous variability in socio-economic, cultural, and health conditions described previously.

When funded insufficiently, however, the results may be worse than those achieved when the federal government was the provider. Because funding of Indian programmes has historically been inadequate, in many instances the programmes taken over by tribes may not have been functioning optimally from the start. Moreover, as is the case with outsourcing in general, transaction costs (e.g. the costs of monitoring contracts) may be high. Further, tribes may not benefit from economies of scale (e.g. with regard to the purchase of goods and services), recruitment and retention of professional staff may be difficult, and tribal and state regulations (for instance with regard to Medicaid) may conflict with one another.

None of these problems is inevitable or insurmountable. Indeed it appears that there is, and will continue to be, great diversity in the funding, functioning and

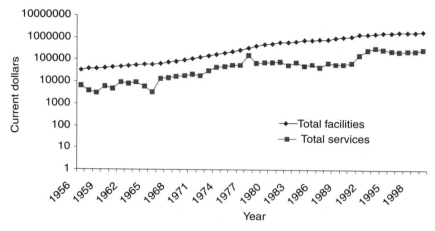

Figure 10.1   Appropriations for Indian health services and facilities, in thousands of current dollars, 1956–99 (semi-log scale).

success of tribally run programmes. Nonetheless, to the degree that budgetary constraints influence all programmes to some degree, serious problems are likely to be more or less widespread and may well have a greater impact on preventive than on clinical programmes (Noren *et al.* 1998; Weiner 1999).

In order to provide a broad picture of the history of funding of health services for Indians, Figure 10.1 displays the total federal appropriations in current dollars since 1956, when the responsibility for health care was shifted from the Bureau of Indian Affairs to the Indian Health Service. There has been a slow increase in the amount available for services each year, with a discernible upward deflection in the 1970s. These increases have not been as rapid as increases in the costs of health care, however.

To account for the increases in the costs of health care, Figure 10.2 displays the amounts available each year from 1990 to 1999, in both current and constant dollars (adjusted for health care costs).

Several points emerge from these data. First, in constant dollars, the budget has remained almost flat over the decade of the 1990s. The increase between 1990 and 1991 is due almost entirely to an increase in the amount available for the construction and maintenance of facilities, not for services. Indeed, as the population is growing more rapidly than the budget for health care, the amount available per capita is actually declining (Noren *et al.* 1999). Secondly, there is a widening gap (in current dollars) between the total budget and the amount appropriated by Congress. This is accounted for by the increasing dependence of the Indian Health Service and Indian tribes which run some or all of their own health services on third party payers such as Medicare, Medicaid, and private insurers (Wellever *et al.* 1998). Moreover, recall that per capita expenditure on Indian health services from all sources is estimated to have been no more than 60–65 per cent of the amount spent on the average U.S. citizen in 1990 (Kunitz 1996). Finally, the income of Indians living on reservations declined in the 1980s (Trosper

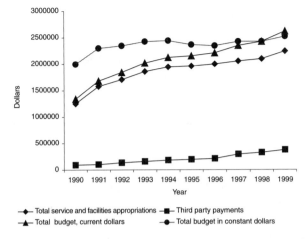

Figure 10.2 Indian Health Service budget, 1990–9, in current and constant dollars.

1996). Although there is no evidence yet available from the 2000 census, it appears unlikely that the trend would have been reversed substantially, if at all, in the 1990s. The available evidence thus suggests that per capita funding for Indian health care is substantially less than for other American citizens and may have decreased during the 1990s; that Indian incomes continue to be substantially lower than those of other citizens; and that after adjusting for the costs of health care, increases in total spending for health services have been minimal, even as new sources of revenue have been accessed.

The new genetic determinism

In this situation of low and stagnating funding, the appeal of genetic explanations for some of the most important chronic diseases has a seductiveness that transcends its very considerable intrinsic interest. The analogy to the germ theory of disease is illuminating. One of the great attractions of the germ theory over the century just past has been the promise of disease control at relatively low cost and without the need for large social interventions. Disease eradication campaigns beginning in the early years of the 20th century had this as an explicit goal (Kunitz 1987). On occasion the promise has been realized, most notably in the case of the vaccine preventable diseases, especially smallpox. On the other hand, many infectious diseases have not responded to this sort of intervention. Thus improvements in living conditions, health care and public health infrastructure, as well as changes in individual behaviour, are still crucial and will become even more so as new and newly important drug resistant infections increase in prevalence. A similar process may be at work with regard to the appeal of genetic explanations of chronic diseases.

In the late 1940s, the Truman Administration attempted to create several federal programmes to improve the health of the American people: national health insurance, federal aid to medical education, federal funding of rural hospitals, and federal support for medical research. In the climate of the McCarthy period and the Cold War, both national health insurance and aid to medical education were defeated. Rural hospital construction and medical research both became federal programmes, and research at the National Institutes of Health began to grow at a dramatic rate. In fact, biomedical research was meant to be a substitute for national health insurance. For if easily applied, economical, technological interventions could reduce the incidence, prevalence, and morbidity from chronic diseases, all could benefit without having to create a national health insurance scheme and/or redistribute income in other ways. Stephen Strickland has written of the 1950s and early 1960s:

> Other avenues – health insurance, hospital care for the aged and indigent, federal aid to medical education – were blocked by a clearly self-serving lobby which was by many leagues more powerful, and which more brazenly fed popular fears, than anything the medical research lobby could hope to be or do. "Medical research," remarked Congressman Laird as he made ready to help boost its budget one more time, "is the best kind of health insurance" the American people could have.
>
> (Strickland 1972)

There have been numerous consequences of these policy choices. One that is especially important is that a formidable research establishment has been created, linking government, industry and university interests. It possesses great sophistication with regard to shaping the nation's research agenda. Much of this agenda requires the expansion of support for genetic research in several fields, including the Human Genome Project (Tauber and Sarker 1993). The expansion and promised benefits of this research have shaped academic medical centres and private corporations and the way scientific careers are made. These institutional and career imperatives have in their turn shaped the way diseases are defined and understood. Edward Yoxen has written:

> [G]iven the intensification of scientific training for medical practice that has occurred, physicians are now much more likely to consider highly technical models of pathological conditions as relevant to their work. Indeed several commentators have drawn attention to the absorption of predominantly white, upper-middle-class physicians and clinical researchers in the United States with the aetiology and molecular genetics of sickle-cell anemia, whilst being relatively unconcerned with the availability of treatment or the delivery of any kind of relevant health care to the poor black people who form the overwhelming majority of sufferers from the disease.
>
> (Yoxen 1982)

This same context has also shaped research interest in the genetic causes of certain non-infectious diseases affecting American Indians.

It is not that biomedical investigators are opposed to minorities receiving appropriate care. Almost certainly just the opposite is true. Nor is the research they do unimportant or uninteresting or of poor quality. On the contrary, much of it is of very high quality and of great interest and importance. Nor was this policy devised with Indians in mind. The growth of the biomedical research establishment was not in lieu of more generous funding for Indian health. Rather, Indians are of interest because they have high rates of some of the diseases that are of particular concern to medical geneticists and molecular biologists, largely because of their high prevalence in the national population. The point is that with funding priorities as they are, the interest of people doing research is almost inevitably drawn to problems that are most relevant to their peers and to funding sources. Generally, this does not include problems of prevention in small and potentially unique populations.

It is important to distinguish among genetic diseases, however. Some single-gene diseases such as certain forms of breast and colon cancer, or certain childhood diseases, that affect Indians may someday soon be amenable to genetic inter-ventions. They are of great importance to affected individuals and families but are not for the most part of great importance in terms of the health of entire populations. The chronic diseases of great public health importance are polygenic in origin (Crawford 1998), to the degree that they are genetic at all. They are thought by many to be a result of the change from the hunting/gathering that characterized the life of our species for most of our existence, to the glut and sedentism of contemporary life in affluent countries (Neel 1962, 1982; Eaton and Konner 1985; Eaton, Konner and Shostak 1988). The conditions in this category that affect Indians most are obesity and non-insulin dependent diabetes (type 2 diabetes) and alcoholism.

With regard to obesity, which is an important risk factor for type 2 dependent diabetes, twin studies have led to estimates that as much as two-thirds of the variance is genetic in origin. But the genetic make-up of the American Indian and total U.S. population did not change over the decade of the 1990s, at a time when the prevalence of obesity among all Americans increased from 12 to 19 per cent. Only a major change in dietary and exercise patterns can explain an increase of such magnitude. George Davey Smith and Shah Ebrahim have written, "The apparent discrepancy between a large amount of variance accounted for by genetic factors and the clear environmental determination of population disease rates" may be accounted for by several factors. Most significant for my purposes is their observation that,

> [T]he contribution of genetic factors to disease rates is an area where Geoffrey Rose's distinction between the determinants of disease rates for a population and factors influencing who gets a disease within a population is crucial. With very general shifts in the population to higher energy intake/energy expen-diture ratio ... the variance between individuals can remain strongly genetically based, but this can make a minimal contribution to the population burden of obesity ...
>
> (Davey Smith and Ebrahim 2001)

The same points can be made about type 2 diabetes. Genetic susceptibility may explain much of the difference between individuals, but it does not explain the rapid increases in prevalence within populations among whom genetic change has not occurred (Szathmary 1994; Young 1994). Moreover, what to do about genetic susceptibility if it exists at higher rates in some populations than others is also highly uncertain. The health care system does not yet have much to offer with regard to the reduction of genetic susceptibility, and perhaps never will (Hall *et al.* 1994).

On the other hand, there is good evidence that social and environmental causes are very important in the aetiology of type 2 diabetes including the intrauterine environment, birth weight, breast versus bottle feeding, and changing dietary and exercise patterns (McDermott 1998). These are all areas in which the health care system can intervene (Wilson *et al.* 1994; Garcia-Smith 1994; Broussard *et al.* 1995). Unfortunately the resources and rewards available to study and implement preventive interventions, especially those targeted to specific populations, do not match those devoted to genetic research, where few if any benefits have yet been realized.

Something similar is true of research on the prevention and treatment of alcoholism in Indian communities. Much has been written on the genetic contribution to alcohol dependence, although evolutionary implications have not been as significant a part of these discussions as they have of diabetes. The co-morbidity of alcohol dependence on the one hand and conduct disorder and anti-social personality disorder on the other has been an important theme in this literature and is useful for illustrative purposes (Petrakis 1985; Lappalianen *et al.* 1998; Slutzske *et al.* 1998; True *et al.* 1999). Many investigators believe that the two latter conditions are genetic in origin and are strongly associated with alcoholism either as the effects of a common genetic cause or as risk factors for alcoholism. Indeed, there is little doubt that conduct disorder and anti-social personality disorder on the one hand and alcoholism on the other are highly correlated and that at the very least the former are important risk factors for the latter. The question is, even assuming that the etiology of conduct disorder and anti-social personality disorder is genetic, at the population level how much alcohol dependence is attributable to these two conditions?

A recent study of an Indian population showed that conduct disorder before age 15 was an important risk factor for alcohol dependence. Not only were people with a history of conduct disorder more likely to become alcohol dependent in adulthood than people without such a history, but they were more severely alcoholic than other alcoholics. On the other hand, the percentage of alcohol dependence attributable to conduct disorder was very small, around 5 per cent (Kunitz and Levy 2000). Thus, even if the alcoholism attributable to conduct disorder were all genetically determined, it would account for very little alcoholism in the population.

This is an example of the prevention paradox described by Geoffrey Rose (1992). That is, if one focuses preventive efforts on high-risk individuals – e.g. children with conduct disorder – they may well benefit from a reduced risk of subsequent alcoholism. At the population level, however, the effects will be virtually imper-

Table10.4 The risk of alcohol dependence and the number of men and women with a lifetime history of alcohol dependence, by conduct disorder score.

| | Women | | | Men | | |
| | Alcohol dependent | | *N* | Alcohol dependent | | *N* |
Score	*n*	%		*n*	%	
0	19	21.1	90	98	55.4	177
1	11	21.2	52	100	75.4	133
2	15	41.7	36	74	71.8	103
3	7	53.8	13	43	81.1	53
4	5	62.5	8	27	93.1	29
5	1	100	1	16	88.9	18
6	1	100	1	9	90.0	10
7	1	50.0	2	7	87.5	8
Total	60	29.5	203	374	70.4	531

Source: Kunitz and Levy (2000).

ceptible because the vast majority of alcoholics do not have histories of conduct disorder. There are many other examples of this same phenomenon: very high serum cholesterol as a risk factor for heart attacks, and severe hypertension as a risk factor for stroke, to list but two. To illustrate, consider Table 10.4.

The data come from the study mentioned above. They indicate that, among both men and women, the greater the severity of the score on a scale measuring conduct disorder before age 15, the greater is the probability of having experienced alcohol dependence in adulthood. They also indicate, however, that the absolute number of alcohol-dependent people is much greater at lower rather than higher levels of conduct disorder. This is the prevention paradox. If efforts are focused on the highest risk people, they will have virtually no effect on the prevalence of alcohol dependence in the population.

What are the implications? Let us assume that both diabetes and alcoholism have large genetic components and that they are polygenic in origin. For population level preventive interventions to be effective, the distribution of risk factors in the entire population must be shifted downward. That cannot be done with regard to genetic risks for chronic disease, for no such genes have yet been identified, and ethical consideration would not permit it even if it were possible. It can be done, however, with diet, with antenatal and well child care, with health education in schools and clinics, and with many other innovative preventive and therapeutic interventions. "The important point is that genes determine *who* may get sick within a class, but environmental factors determine the *frequency* of sickness among susceptibles" (Baird 1994). It is clear, however, from the ratios of death rates due to amenable causes displayed in Table 10.3 that too little intervention in these environmental factors is being done to reduce the burden of these conditions.

Conclusions

I have suggested that, along with continuing poverty and unemployment, part of the reason for continuing disparities between the health of Indians and non-Indians in the United States is budget constraints. There is simply insufficient money to permit the level of preventive public health interventions that is required to have a measurable impact. In this context, it is not at all certain that devolution to tribal governments of responsibility for services will have any more profound impact on these discrepant rates than the Indian Health Service had when it was fully responsible for the provision of population and personal services.

Exacerbating this problem is the fact that so many resources, and such high value, have been placed on the promise of genetic research for improving the health of populations, particularly with regard to chronic diseases. Biomedical research was from its inception after World War II understood by policy makers to be an alternative to the universal provision of health care to the American people. This same understanding has helped fuel the continued growth of the National Institutes of Health and the research it has sponsored into the genetic causes of chronic diseases.

The problem is not that research into the genetic contribution to human diseases is a bad idea, or that there are not more or less significant genetic causes of many conditions, or that other kinds of research have been entirely ignored. The problem is that too much has been expected from biomedical research. It has been expected to produce technological innovations that will make national health insurance unnecessary.

The consequences have been numerous and also paradoxical. (1) Biomedical innovations that have been developed have often as not increased costs and exacerbated rather than reduced inequality in access to, and utilization of, health care. This is because many of them are half-way technologies, not definitive, inexpensive technologies that can eliminate diseases as vaccines can (Thomas 1974). (2) The most rewarding investigative careers with the greatest possibilities of funding are likely to be made in genetic and molecular biological research. This further shapes the ways in which disease etiology is conceived and has contributed to a growing sense of genetic predestination. In contrast to predestination, genetic predisposition to develop a particular disease implies probability and risk rather than certainty. But in social policy as well as in professional and popular discourse, including among many Indians, predisposition is often conflated with predestination, so that a statistical correlation becomes a cause: "A genetic predisposition to alcoholism, for example, becomes an 'alcoholic gene'." (Nelkin and Lindee 1995). And finally, (3) research on population-based preventive interventions has received less support than it might have otherwise.

All of these national developments have had an important impact on Native Americans, for Native Americans manifest high rates of some of the conditions that are of most interest to investigators. Thus they are highly likely to be asked to be the subjects of genetic research that has little likelihood of benefiting them, while the kind of research that might lead to useful preventive interventions has received far less support than it deserves. This is especially important since Indian populations differ in many ways that may make different approaches to the same problem, whether alcohol dependence or diabetes, appropriate (Broussard *et al.* 1974; Levy and Kunitz 1974). It is not immediately obvious that Indian health workers know any better than anyone else how to prevent these conditions. That is why studies of different preventive interventions are needed. In the face of constrained budgets and a prevailing ideology of genetic predisposition if not predestination, however, devolution of control of services to tribal governments, which ideally should result in locally appropriate preventive interventions, is unlikely to have such hoped for beneficial results.

The answer is not to reduce or discontinue genetic research. Individual tribes will decide for themselves whether they wish to participate or not, and what the conditions of participation will be. Nor is the answer to reverse the policy of devolution. That again is a tribal decision, and different tribes have made, and will continue to make, different decisions. To adequately begin to deal with the chronic disease problems that increasingly afflict Native Americans, adequate resources and incentives, both financial and professional, need to be available to test and implement a variety of prevention programmes that address these issues in different tribal populations.

Acknowledgements

Jerrold E. Levy and T. Kue Young commented on an early version of this chapter.

References

Adams, A. (2000) 'The road not taken: how tribes choose between tribal and Indian Health Service management of health care resources', *American Indian Culture and Research Journal*, 24: 21–38.

Baird, P.A. (1994) 'The role of genetics in population health', in R.G. Evans, M.L. Barer and T.R. Marmor (eds) *Why Are Some People Healthy and Others Not? The Determinants of Health of Populations*, New York: Aldine De Gruyter, pp. 133–59.

Bergman, A.B., Grossman, D.C., Erdrich, A.M., Todd, J.G. and Forquera, J. (1999) 'A political history of the Indian Health Service', *The Milbank Quarterly*, 77: 571–604.

Broussard, B.A., Sugarman, J.R., Bachman-Carter, K., Booth, K., Stephenson, L., Strauss, K. and Gohdes, D. (1995) 'Toward comprehensive obesity prevention programs in Native American communities', *Obesity Research* 3, supp. 2: 289s–97s.

Crawford, M. (1998) *The Origins of Native Americans: Evidence from Anthropological Genetics*, Cambridge: Cambridge University Press.

Cunningham, P.J. (1996) 'Healthcare utilization, expenditures, and insurance coverage for American Indians and Alaska Natives eligible for the Indian Health Service', in G.D. Sandefur, R.R. Rindfuss and B. Cohen (eds) *Changing Numbers, Changing Needs: American Indian Demography and Public Health*, Washington, DC: National Academy Press, pp. 289–314.

Davey Smith, G. and Ebrahim, S. (2001) 'Epidemiology – is it time to call it a day?', *International Journal of Epidemiology*, 30: 1–11.

Eaton, S.B., Konner, M. and Shostak, M. (1988) 'Stoneagers in the fast lane: chronic degenerative diseases in evolutionary perspective', *American Journal of Medicine*, 84: 739–49.

Eaton, S.B. and Konner, M. (1985) 'Paleolithic nutrition: a consideration of its nature and current implications', *New England Journal of Medicine*, 312: 283–9.

Garcia-Smith, D. (1994) 'The Gila River diabetes prevention model', in J.R. Joe and R.S. Young (eds) *Diabetes as a Disease of Civilization: The Impact of Culture Change on Indigenous Peoples*, New York: Mouton de Gruyter, pp. 471–503.

Hall, T.R., Hickey, M.E. and Young, T.B. (1994) 'Many farms revisited: evidence of increasing weight and non-insulin dependent diabetes in a Navajo community', in J.R. Joe and R.S. Young (eds) *Diabetes as a Disease of Civilization: The Impact of Culture Change on Indigenous Peoples*, New York: Mouton de Gruyter, pp. 129–46.

Holland, W.W. (ed.) (1993) *European Community Atlas of Avoidable Death*, vol. 2, Oxford: Oxford University Press, p. 1.

IHS (1998–99) *Regional Differences in Indian Health 1998–99*, Rockville, MD: Department of Health and Human Services, Public Health Service, Indian Health Service, p. 104.

Keane, C., Marx, J. and Ricci, E. (2001a) 'Perceived outcomes of public health privatization: a national survey of local health department directors', *Milbank Quarterly*, 79: 115–37.

Keane, C., Marx, J. and Ricci, E. (2001b) 'Privatization and the scope of public health: a national survey of local health department directors', *American Journal of Public Health*, 91: 611–17.

Kunitz, S.J. (1983) *Disease Change and the Role of Medicine: The Navajo Experience*, Berkeley: University of California Press.

Kunitz, S.J. (1987) 'Explanations and ideologies of mortality patterns', *Population and Development Review*, 13: 379–408.

Kunitz, S.J. (1994) *Disease and Social Diversity: The Impact of Europeans on the Health of Non-Europeans*, New York: Oxford University Press.

Kunitz, S.J. (1996) 'The history and politics of US health care policy for American Indians and Alaskan natives', *American Journal of Public Health*, 86: 1464–73.

Kunitz, S.J. and Levy, J.E. (2000) *Drinking, Conduct Disorder, and Social Change: Navajo Experiences*, New York: Oxford University Press.

Lappalainen, J., Long, J.C., Eggert, M., Ozaki, N., Robin, R.W., Brown, G.L., Naukkarinen, H., Virkkunen, M., Linnoila, M. and Goldman, D. (1998) 'Linkage of antisocial alcoholism to the serotonin 5-HT1B receptor gene in two populations', *Archives of General Psychiatry* 55: 989–94.

Levy, J.E. and Kunitz, S.J. (1971) 'Indian reservations, anomie, and social pathologies', *Southwestern Journal of Anthropology*, 27: 97–128.

Levy, J.E. and Kunitz, S.J. (1974) *Indian Drinking: Navajo Practices and Anglo-American Theories*, New York: John Wiley and Sons.

May, P.A. (1986) 'Alcohol and drug misuse prevention programs for American Indians: needs and opportunities.' *Journal of Studies on Alcohol*, 47: 187–95.

McDermott, R. (1998). 'Ethics, epidemiology and the thrifty gene: biological determinism as a health hazard', *Social Science and Medicine*, 47: 1189–95.

Neel, J.V. (1962) 'Diabetes mellitus: a 'thrifty' genotype rendered determental by "progress"?', *American Journal of Human Genetics*, 14: 353–62.

Neel, J.V. (1982) 'The thrifty genotype revisited', in J. Koeblerling and R. Tattersall (eds) *The Genetics of Diabetes Mellitus*, Serono Symposium, no. 47, New York: Academic Press.

Nelkin, D. and Lindee, M.S. (1995) *The DNA Mystique: The Gene as a Cultural Icon*, New York: W.H. Freeman, p. 166.

Noren, J., Kindig, D. and Sprenger, A. (1998) 'Challenges to Native American health care', *Public Health Reports*, 113: 22–33.

Omran, A.R. (1971) 'The epidemiologic transition: a theory of the epidemiology of population change', *Milbank Memorial Fund Quarterly*, 49: 509–38.

Petrakis, P.L. (1985) *Alcoholism: An Inherited Disease*, U.S. Department of Health and Human Services, Public Health Service, Alcohol, Drug Abuse, and Mental Health Administration, National Institute on Alcohol Abuse and Alcoholism, Washington, DC: U.S. Government Printing Office.

Rose, G. (1992) *The Strategy of Preventive Medicine*, Oxford: Oxford University Press.

Rutstein, D.D., Berenberg, W., Chalmers, T.C., Child, C.G., Fishman, A.P. and Perrin, E.B. (1976) 'Measuring the quality of medical care: a clinical method', *New England Journal of Medicine*, 294: 582–8.

Shoemaker, N. (1999) *American Indian Population Recovery in the Twentieth Century*, Albuquerque: University of New Mexico Press.

Slutzske, W.S., Heath, A.C., Dinwiddie, S.H., Madden, P.A., Bucholz, K.K., Dunne, M.P., Statham, D.J. and Martin, N.G. (1998) 'Common genetic risk factors for conduct disorder and alcohol dependence', *Journal of Abnormal Psychology*, 107: 363–74.

Strickland, S. (1972) *Politics, Science, and Dread Disease*, Cambridge: Harvard University Press, p. 213.

Szathmary, E. (1994) 'Factors that influence the onset of diabetes in Dogrib Indians of the Canadian Northwest Territories', in J.R. Joe and R.S. Young (eds) *Diabetes as a Disease of Civilization: The Impact of Culture Change on Indigenous Peoples*, New York: Mouton de Gruyter, pp. 229–68.

Tauber, A.I. and Sarkar, S. (1993) 'The ideology of the human genome project', *Journal of the Royal Society of Medicine*, 86: 537–40.

Thomas, L. (1974) 'The technology of medicine', in L. Thomas (ed.) *The lives of a cell: notes of a biology watcher*, New York: Viking Press, pp. 31–6.

Trosper, R.L. (1996) 'American Indian poverty on reservations, 1969–1989', in G.B. Sandefur, R.R. Rindfuss and B. Cohen (eds) *Changing Numbers, Changing Needs: American Indian Demography and Public Health*, Washington, DC: National Academy Press, pp. 172–95.

True, W.R., Heath, A.C., Scherrer, J.F., Xian, H., Lin, N., Eisen, S.A., Lyons, M.J., Goldberg, J. and Tsuang, M.T. (1999) 'Interrelationship of genetic and environmental influences on conduct disorder and alcohol and marijuana dependence symptoms', *American Journal of Medical Genetics*, 88: 391–7.

Verrano, J.W. and Ubelaker, D.H. (1992) *Disease and Demography in the Americas*, Washington, DC: Smithsonian Institution Press.

Weiner, D. (1999) 'Ethnogenetics: interpreting ideas about diabetes and inheritance', *American Indian Culture and Research Journal*, 23: 155–84.

Wellever, A., Hill, G. and Casey, M. (1998) 'Commentary: Medicaid reform issues affecting the Indian health care system', *American Journal of Public Health*, 88: 193–5.

Wilson, R., Graham, C., Booth, K.G. and Gohdes, D. (1994) 'Community approaches to diabetes prevention', in J.R. Joe and R.S. Young (eds) *Diabetes as a Disease of Civilization: The Impact of Culture Change on Indigenous Peoples*, New York: Mouton de Gruyter, pp. 495–503.

Young, T.K. (1994) *The Health of Native Americans: Towards a Biocultural Epidemiology*, New York: Oxford University Press.

Yoxen, E. (1982) 'Constructing genetic diseases' in P. Wright and A. Treacher (eds) *The Problem of Medical Knowledge: Examining the Social Construction of Medicine*, Edinburgh: Edinburgh University Press, pp. 141–61.

11 The public's health

The changing role of public health

Jean Peters

Introduction

Public health is the science and art of promoting, protecting, and improving health and well-being through organised efforts of society (Department of Health 1988). This chapter provides an overview of how the determinants of health, and consequently disease profiles, have changed over time in terms of emphasis and influence, and what role public health has, and can play, sometimes with government intervention, to maximise the public's health given these changes.

Health and disease can be considered as two extremes of a continuum. However, since in general it has proved difficult to produce a definition of 'health' with measurable outcomes, biologists, anthropologists, health care professionals, and researchers have tended to focus on factors that indicate absence of health, ill-health, degrees of ill-health, or actual disease state. The classic definition of health is that proposed by the World Health Organisation in 1948: 'Health is a state of complete physical, mental and social well-being and not merely the absence of disease'. This definition, however, is not very helpful as not only does it describe an ideal state rarely attained in the real world, it also does not provide any clear measurable factors with which to assess if such a state has been achieved. The definition of Stokes *et al.* (1982) is a more helpful one as it acknowledges those aspects of health that can be measured: 'Health is a state characterised by anatomical integrity, ability to perform personally valued family, work and community roles; ability to deal with physical, biologic, and social stress; a feeling of well-being; and freedom from the risk of disease and ultimately death'. From this definition it can be inferred that disease compromises anatomical integrity, limits performance, can be physically or mentally impeding and possibly fatal, all factors with measurable outcomes. Being able to measure health or disease is important if health is seen as important among the objectives and values of most individuals, and such individuals expect governments and administrations to pursue policies that will give them the opportunity to live a healthy life of optimum duration and quality. Alternatively, or in addition, the governments and other administrations themselves regard attainment of health or maintenance of a healthy population as important, for whatever reason.

Historical perspective

In previous centuries, morbidity and mortality rates were high and predominantly attributable to infectious diseases. Plagues swept Europe in the seventeenth and eighteenth centuries in pandemics every ten years or so, carried by rats and transmitted from one infected person to another by droplet contamination through sneezing and coughing. The spread of the disease was exacerbated by lack of knowledge of its transmission.

The industrial revolution of the eighteenth and early nineteenth centuries also provided an environment that exacerbated transmission of infectious diseases. With the massive migration of populations to the cities, large quantities of housing were built but levels of sanitation were generally poor and overcrowding was endemic. It was during this time that the modern public health movement began with the work of reformers such as Edwin Chadwick, a lawyer interested in the living conditions of the poor and the relationship between chronic sickness and poverty. The inquiries that he, and a contemporary, Dr Southwood Smith, instigated and published around this issue prompted the Government to set up a Royal Commission, one outcome of which was the appointment of the first Medical Officer for Health in 1847 in Liverpool. An outbreak of cholera, also in 1847, prompted the Government to pass the Public Health Act in 1848 and to set up a General Board of Health. In a further cholera epidemic in London in 1854, John Snow, applying epidemiological principles, was able to demonstrate that the cholera was spread by an invisible agent in water, knowledge of the cholera bacterium remaining unknown for several more years. By monitoring and mapping out cases, he was able to identify the disease transmission pattern and thus identify a potential source. An intervention, removal of the handle of a pump, resulted in control of the infection.

Whilst mortality was predominantly due to infectious disease, diseases and ill-health not attributable to infection were also present in the population. For example, data on mortality collected by John Graunt (1620–74) in his table of diseases and casualties, 1632, included that on accidents, deliberate or otherwise.

Changing patterns of disease

Between the eighteenth and nineteenth century, mortality rates started to fall with 75 per cent of the improvement seen from 1848–54 to 1971 being attributable to a decline in infectious diseases (Gray and Payne 1993). Non-infectious conditions accounted for the remaining decline, although for some specific conditions increases in death rates occurred. The non-infectious diseases that dominate the morbidity and mortality statistics today, whilst present in previous centuries, had less of an impact then because infectious diseases were killing many in early life, leaving few to survive to an age at which the non-infectious diseases, such as cardiovascular disease and cancers, have an impact. The decline in overall mortality rate has not been linear over the 120-year period. Seventy per cent of this decline has occurred in the last 70 years, 30 per cent in the first 50 years; during the same time the ratio of infectious to non-infectious disease changed from 11:1 to 2:1. Infectious disease, which accounted for 25 per cent of all deaths in the nineteenth century, accounted for only 1 per cent in the twentieth century (Gray and Payne 1993).

At least some of this change in the disease profile can be attributed to an early public health approach. John Snow's demonstration of transmission of infection, and its prevention, combined with the powers of the Public Health Act and of local Medical Officers for Health enabled local authorities to take control of, and set standards for environmental factors, such as water supplies and methods for disposal of sewage. By the end of the nineteenth century, a central government department of public health and local departments of public health had been established in every local government district. The school health service was set up in 1907, ante- and post-natal care services in 1915, and in the 1920s and 1930s the Medical Officer for Health was responsible for public provision of health care, the monitoring of water supplies, sewage disposal, food hygiene, housing, and the control of infectious diseases (Department of Health 2001a). The Peckham experiment in 1935 provided health and social services for those too poor to purchase medical help.

As a net result, there have been a number of significant achievements and improvements in population health and reduction of disease this century. As summarised by Lee (2001), vaccination has brought about the deliberate extinction of the smallpox micro-organism, control of measles, rubella, diphtheria, polio, tetanus, and tuberculosis. Cholera and typhoid have been more or less eliminated with a safe water supply. Antibiotics treat and cure established infections due to these and many other organisms. Along with better water supplies and waste disposal are improvements in general living conditions at home and at work, reducing the impact of transmissible diseases and work-related illnesses. Health and safety regulations have helped to reduce job-related injuries. Care of mothers and babies before, during and after birth has reduced prenatal and maternal mortality through improvements in hygiene, nutrition, logistics and therapy. Family planning supports choice in the numbers and the timing of births. Nutritional improvements have virtually eliminated goitre, rickets and pellagra.

In spite of these benefits, inequalities in health and differences in prevalence, incidence and mortality of certain diseases within and between populations have not disappeared. In 1974, the Medical Officer for Health post was abolished and the outcome of an enquiry, *Public Health in England*, reported that public health had lost its way (Department of Health 1988). Following the recommendations of that report, health authorities were given responsibility for assessing the health needs of their populations, of allocating resources according to identified need, and of evaluating outcome. They were also required to appoint a director of public health. Since then public health practitioners have played a greater role in reshaping health services and in the development of evidence-based health care within the NHS.

Changes in thinking

Changes in thinking about health and disease have also contributed to the approach to disease management. Until around the eighteenth century, humoral theories of illness dominated clinical practice, but as knowledge increased, thinking changed and illnesses began to be seen as discrete pathological lesions and a biomedical model of disease evolved (Armstrong 1989). This model has been very successful in accounting for, and explaining, many conditions but by its very success it has tended to mask alternative explanations and understandings of the nature of illness such as social influences on health (Armstrong 1989). A recent approach has been to look beyond or away from the strict biomedical model and to consider the longer-term social and ethical issues. In terms of public health, the medical model's emphasis is on prevention of specific diseases through medical interventions such as screening, immunisation, or disease cure and control, whilst the social model gives precedence to the wider determinants of health, factors such as employment, social class, and education. The Public Health Green paper *Reducing health inequalities: an action report* (Department of Health 1999a) embraces both of these, categorising factors affecting health as:

- Fixed: genes, sex, ageing
- Social and economic: poverty, employment, social exclusion
- Environment: air quality, housing, water quality, social environment
- Lifestyle: diet, physical activity, smoking, alcohol, sexual behaviour, drugs
- Access to services: education, NHS, social services, transport, leisure.

In any population, for any health condition, biomedical aspects (biology, pathology and genetics) are interwoven with psychosocial and economic factors. Thus, whilst contaminated water delivered by a water pump was essential for the cholera to occur in 1854, factors other than infection, such as the living conditions and practices of the household members who used the water pump, also had an impact on the levels of sickness and death seen.

Changing lifestyles

So whilst the biomedical model can be used in part to explain the emergence, decline, and re-emergence of diseases, such changes are also a response to changes in lifestyle, environment and socio-economic circumstances. It has been argued that the large decline seen in tuberculosis mortality from the middle of the nineteenth century (when causes of deaths were first recorded) until recently is attributable predominantly to improvements in nutrition and the standard of living, not medical intervention. The tuberculosis bacillus was discovered in the late 1800s, and a specific treatment (streptomycin) for it was discovered in 1948, but by 1948 over 90 per cent of the decline in mortality from the disease had already occurred (Gray and Payne 1993). Nutritional improvements may be similarly responsible for the decline seen in a number of other conditions with medical interventions estimated to account for only about 4 per cent of the total improvement in life expectancy (McKeown 1979). However the theory has its limitations, for example, smallpox elimination has been attributed to a vigorous vaccination programme and tuberculosis is a re-emerging problem, particularly in people with compromised immune systems. Its previous decline may not have been attributable to improved living standards but to a decline in other debilitating diseases, responding to either, or both, biomedical and social interventions, thus leaving the individual potentially susceptible to tuberculosis.

Irrespective of the actual balance between the biomedical and socio-economic, we are now seeing domination in the twenty-first century of the so-called lifestyle diseases, the cancers, cardiovascular diseases, and type 2 diabetes, all of which have associations with one or more behaviours such as smoking, overeating, and insufficient exercise. For example, data on body mass index demonstrate an upward trend from 23.8 in 1980 to 26.4 in 1999 in women. With the increased risk of type 2 diabetes with increasing body mass index (Colditz *et al.* 1995) prevalence of type 2 diabetes is predicted to increase by just under 100 per cent in 25 years, from 1980 to 2005. Furthermore, this risk of body mass index related type 2 diabetes is present in younger as well the more usual older age groups and there is a strong ethnic component.

Changing perceptions of responsibility

In the 1980s and early 1990s there was a perception that it was an individual's fault if they were suffering from, or at risk of, a so-called lifestyle disease as they chose the lifestyle they adopted. Thus individuals had responsibility for their own health and behaviour. The Health of the Nation strategy (Department of Health 1992) identified five key health areas associated with lifestyle behaviours: coronary heart disease, accidents, mental health, sexual health, and cancers, and set targets for reductions in prevalence or incidence of each condition or in behaviour activity, such as smoking, associated with each. The Health of the Nation strategy also defined the aim of health education as 'to ensure that individuals are able to exercise informed choice when selecting the lifestyle which they adopt'. The Health of the Nation approach failed. Evaluation showed that merely having targets was insufficient to ensure action and that a biomedical model approach was inappropriate. A strategy centred in the health service and relying on medical expertise to lead its implementation and evaluation did not address all the relevant issues. The roles of public health in its broadest sense were marginalised, as were key institutions such as local authorities and environmental health departments (Department of Health 1998a).

In theory, identification of disease risk factors means individuals can make informed lifestyle choices. However an individual's health is not totally within the control of that individual. Human behaviour does not reflect individual choices alone but is influenced by the social, economic and political environments, and these are usually beyond the control of the individuals who are affected by them. *Saving lives: our healthier nation* (Department of Health 1999b) adopted a philosophy in which social, economic and environmental factors were seen as important for health and an approach in which people, communities and Government needed to work together in partnership to improve health. With this strategy, targets were again set covering four areas, cancer, coronary heart disease and stroke, accidents, and mental illness, with reductions to be achieved in incidence and prevalence of each by 2010 but using a different approach to that of the Health of the Nation (Department of Health 1992).

One important issue is the balance between individual versus state responsibility. Both are needed for health to improve, e.g. it is an individual's option to smoke, or not to smoke, a government's option to tax or not to tax cigarettes. Traditionally governments have only accepted limited responsibility for the health and welfare of an individual but not necessarily given society any input into its decision-making. Members of society, the patients or consumers, need to be allowed to have involvement in their own health and healthcare, and local norms and cultures must be considered in any debate and action. The following research illustrates a partnership approach taken between public health practitioners, local residents and environmental health officials. It involved responding to public concern, monitoring the problem, and lobbying where necessary and for appropriate policy changes to reduce risk. Following concern expressed by residents in one geographical area of Hong Kong about the impact of locally recorded high levels of air pollutants on their children's respiratory health, an epidemiological study was carried out by public health professionals. The resulting evidence, which demonstrated an excess risk of respiratory symptoms in children living in the air polluted area, when presented to the Environmental Protection Department, a government body, contributed to a policy change that set lower maximum permitted levels for sulphur in fuel. The resulting 84 per cent fall in ambient levels of sulphur dioxide in that area of the territory, an area with high levels of industry, was followed by a reduction in risk of respiratory symptoms in the local children to a level comparable with others living in areas of the territory where sulphur dioxide levels had always been much lower (Peters *et al.* 1996) (Table 11.1).

Unchanging inequalities in health

Irrespective of where the balance of control does and should lie, an imbalance with respect to health and disease across the population continues to exist in spite of any overall improvements. Life expectancy in England and Wales was 52 years for men, and 55 for women in 1910. In 1994, whilst it had improved, the male:female difference remained at 74 and 79 years respectively (Department of Health 2001a). Similarly, standardised mortality ratios in England and Wales in 1930/1932 were 1.2 times higher in the unskilled compared with professional groups, by 1991/1993 the difference was 2.9 times. With respect to ethnicity, mortality ratios ranged from 52 for those of West African origin in England and Wales in 1989/1992 to 150 for those from South Asia. The gap between rich and poor widens further when life expectancy is divided into years in good health and years of disability. The poor not only have shorter lives than the non-poor, but a greater part of their lifetime is spent with disability. Whilst overall, people of all social classes have greater wealth and population health indicators have improved, health inequalities between members of society have grown over recent decades. Circumstances are not necessarily worse for those at the bottom of the social class

Table 11.1 Selected respiratory symptoms, adjusted for confounding factors, before and after the air pollution intervention, odds ratios (OR) with 95% confidence intervals (CI)

Variable		Before air pollution intervention 1989/90			After air pollution intervention 1991		
		OR	95% CI	Sig	OR	95% CI	Sig
Cough or sore throat	District (polluted vs less polluted)	1.22	1.05–1.42	<0.05	0.92	0.73–1.15	≥0.05
	Sex (boy vs girl)	1.35	1.17–1.56	<0.0001	1.59	1.27–1.99	<0.0001
	Age (years)	0.83	0.78–0.89	<0.0001	0.79	0.63–0.85	<0.0001
	Smokers 1 vs 0	1.20	1.04–1.40	<0.05	1.23	0.97–1.56	≥0.05
	2–4 vs 0	1.67	1.36–2.06	<0.0001	1.55	1.09–2.21	<0.05
Phlegm	District (polluted vs less polluted)	1.11	0.96–1.30	≥0.05	0.88	0.68–1.13	≥0.05
	Sex (boy vs girl)	1.33	1.15–1.54	<0.001	1.41	1.10–1.81	<0.01
	Age (years)	0.79	0.74–0.85	<0.0001	0.86	0.73–1.01	
	Smokers 1 vs 0	1.33	1.15–1.54	<0.001	1.26	0.96–1.64	
	2–4 vs 0	1.95	1.58–2.40	<0.0001	1.75	1.19–2.56	<0.01
Wheeze or asthma	District (polluted vs less polluted)	1.27	1.04–1.54	<0.05	1.14	0.89–1.47	≥0.05
	Sex (boy vs girl)	2.11	1.73–2.56	<0.0001	2.74	2.09–3.59	<0.0001
	Age (years)	0.91	0.85–0.98	<0.001	0.81	0.70–0.93	<0.01
	Smokers 1 vs 0	1.02	0.85–1.22	≥0.05	0.91	0.69–1.19	≥0.05
	2–4 vs 0	1.11	0.85–1.46	≥0.05	1.55	1.08–2.23	<0.05

Notes
Other variables: session of study, housing, father's education were adjusted in the model, results not shown.

grades, they are much better for those at the top, the health of the poor is improving, but this is at a slower rate compared with that seen in the better off. The Select Committee on Health (Department of Health 2001a) reported that if all infants and children up to age 15 enjoyed the same survival chances as those children from social classes I and II, over 3000 deaths a year might be prevented. For adults, bringing all aged 16–64 up to the mortality experience of social class I would mean 39,000 fewer deaths per year. A number of deprivation indices have been developed, for example, Townsend (Townsend *et al.* 1988), in an attempt to provide a measure of inequality. These indices are an aggregation, sometimes weighted, of a number of societal variables including one or more of the following: car ownership, type of housing, owner/occupier status, social class, employment status, crowding in the home, rural/urban location.

Poor health is not necessarily an outcome solely of absolute poverty but also of relative poverty – one becomes unhealthy if one perceives oneself to be poorer than others (Wilkinson 1999). The health statistics suggest that health is predicated on an individual's position in society at every level. A large Whitehall study of civil servants has provided some understanding of the relationship between social gradients and health in an organisation where both the richest and poorest groups are excluded, and jobs are ascribed to grades in relation to seniority. In this population it was found that a person's position in the hierarchy was related to mortality, coronary heart disease, cancers, and accidents, suggesting that those who have low control over their work environment are at higher risk of disease than people who had high control (Bosma *et al.* 1997).

Preventing disease

The evidence on health inequalities and inequalities in health care services, their delivery or access, is plentiful (Department of Health 1998b). The key issue is what to do to address and eliminate the problem. The World Health Organisation asserts that the differing degrees of efficiency with which health systems organise and finance themselves, and react to the needs of their populations, explains much of the widening gap between the rich and poor, in countries and between countries around the world (WHO 2001). How health systems, and the estimated 35 million or more people they employ world-wide, perform makes a profound difference to the quality and value, as well as the length, of the lives of the billions of people they serve. An emerging role for public health in the last half century has been that of understanding and addressing inequalities of use, need and access in health systems and services.

The ultimate responsibility for overall performance of a country's health system lies with government, which in turn should involve all sectors of society in its stewardship. But consumers also need to be better informed about what is good and bad for their health, why not all of their expectations can be met, and that they have rights which all providers should respect (WHO 2001). Evidence on the effectiveness of specific interventions to tackle health inequalities is limited. Projects that improve an individual's job and education prospects, or incomes and educational standards in society do not necessarily work; the focus may need to be on reducing status differences which drive the competitive element in consumption (Wilkinson 1999).

Even with the many initiatives generated, uptake is often better by consumers in the advantaged sectors, thus exacerbating the gap. If the focus is specifically on deprived groups, the bottom 10 per cent for example, numbers are often too small to have any impact. To maximise impact, interventions need to target the bottom 20–40 per cent with a sliding scale of input (Department of Health 2001a).

One form of intervention and a key role for public health is that of prevention of disease and ill-health. Prevention of a condition such as coronary heart disease can make a bigger impact on a population's health than treating, but not necessarily curing, those who have succumbed to the condition. Unfortunately this proposition is not self-evident since diseases that are prevented are necessarily unreported and therefore the success of preventive therapy is more difficult to demonstrate than, for example, drug therapy. It is also difficult to demonstrate that, for those who do not get the disease, prevention strategies can take the credit. Some preventive strategies can bring large benefits to a community but they may bring little to individuals, and only to a small fraction of those exposed to the strategy. One key decision with prevention strategies is who to target, those at high risk or the whole population. The findings from the Hong Kong study illustrate the debate of a targeted versus blanket intervention approach. Policy changes which covered the whole territory, and which in the case of the air quality intervention cost HK$64bn (Barron *et al.* 1995), did not necessarily change status or provide benefit for those living in areas where air quality was not compromised. Conversely, the same epidemiological study demonstrated excess risks of respiratory symptoms in children living with smokers, after adjustment for the district (air quality) effect. These risks for some symptoms were higher and did not reduce following the policy change (Table 11.1) (Peters *et al.* 1996). In terms of reducing respiratory disease, it could have been cheaper to target smokers, approximately 40 per cent of men and 3 per cent of women in the territory, although not necessarily easier.

Changing government approach

Improvements in public health depend upon the importance attached to the public's health by policy makers both within and outside the health service. One of the problems in the health debate at present is that health and health care are dominated by acute care. The *NHS Plan* (Department of Health 2000a) has eclipsed the public health approach championed in *Saving Lives: Our Healthier Nation* (Department of Health 1999b). This may be because the mindset of the NHS is not allowing public health to move far beyond the medical model (Department of Health 2001a). Public health appeared to be a Government priority with the appointment of a minister for public health and high profile publications, such as the Acheson report on inequalities (Department of Health 1998b), *Reducing Health Inequalities* (Department of Health 1999a), *Saving Lives: Our Healthier Nation* (Department of Health 1999b), and *Strengthening the Public Health Function* (Department of Health 2001b).

The Select Committee on Health (Department of Health 2001a) saw great potential for public health to strengthen its approach with respect to health inequalities through the director of public health role and local and health authorities. However, the imminent changes (in April 2002) with closure of district health authorities in England and the movement of the public health function into primary care groups and trusts, in conjunction with local authorities, is likely to impede progress, albeit hopefully only temporarily.

Attempts made by those working in public health in partnership with other organisations to improve population health and reduce disease are also being disrupted by a plethora of new initiatives, such as: Health Action Zones, Sure Start, Sure Start Plus, Education Action Zones, regeneration, healthy living centres, fuel poverty strategies, tax credit systems, health improvement plans, beacons and beacon status, NHS Direct, and Public Health Observatories. For effective outcome, many of these interventions require partnership, joined up links and co-ordination between different government departments, statutory agencies, elected authorities and the voluntary sector, but these take time to establish. For many of these initiatives there is also a requirement by Government for 'early wins'. In the case of Health Action Zones, there was a Government-imposed change in focus partway through the initiative, to a greater focus on cardiovascular disease and secondary care interventions rather than on community-led innovative initiatives, although this has not necessarily limited the health inequalities agenda (Peters *et al.* 2002). New initiatives are being presented before earlier programmes have been in place long enough to be evaluated and any evaluation in the short-term has only been able to focus on process rather than health outcomes. Currently, national service frameworks are being launched for specific diseases and age-groups with standards to be set in place over a given timescale. However the focus on the public's health in terms of prevention of the condition, a key role for public health, appears to have been marginalised. In the coronary heart disease national service framework, only two of the 12 standards, little of the money, and none of the infrastructure relate to primary prevention (Department of Health 2000b).

Conclusions

Is the role of public health changing? What changes is the degree of concentration on different aspects of the public health function and the requirement for different emphases at different times. There is a web of influences, biomedical and socio-economic, self-inflicted or externally driven, that interrelate with each other. At any one point in time some are rising in importance whilst others decline, but all have an impact, to a lesser or greater extent, on the level and type of disease seen in society. The core components of public health, the monitoring, the observation are always needed, and the Public Health Observatories, launched in 2000, provide the modern technological input to this function. The traditional epidemiological approach used by John Snow to map and trace contacts in the cholera epidemic in 1854 is still used today in communicable disease control.

So where are we now? Patterns of disease have changed over the past two hundred years and are still evolving. The Government's approach and that of public health continues to change and has not always worked in favour of health improvements. The government's and populations' perceptions of health have changed, as have people's expectations of their own health and what can be done to treat or cure any illness, even if erroneous. Some actions to improve health have generated other health problems and future actions must consider all factors, especially the longer-term outcomes. In the past, residents from slum clearance left overcrowded tenements with no indoor sanitation or heating, but they moved into housing that soon became associated with other problems, crime, vandalism, and social isolation. The focus on the environmental aspects ignored the social factors. There is emerging evidence that very hygienic practices and lack of exposure to infection may be responsible for the increase in prevalence of asthma (von Mutius 2001).

Can public health deliver society's expectations to be healthy? It requires public health to keep the wider determinants of health on governments' and society's agendas and to promote changes in population norms of behaviour and risk perception. In addition, there are challenges to the public health role through demographic change, the development of new medical technologies and procedures, the emergence of new diseases and new threats to public health, and rising aspirations of the public for a better life. Public health needs to maintain the desire, the zeal and conviction of the earlier social reformers that things could be better and that the public's health could and should be improved. This societal change is needed alongside medical and scientific progress coupled with appropriate action from professionals, policymakers and the public themselves.

References

Armstrong, D. (1989) *An Outline of Sociology as Applied to Medicine*, Cambridge: Butterworth Heinemann.

Barron, W.F., Liu, J., Lam, T.H., Wong, C.M., Peters, J. and Hedley, A.J. (1995) 'Costs and benefits of air quality improvement in Hong Kong', *Contemporary Economic Policy*, 13: 105–17.

Bosma, H., Marmot, M.G., Hemingway, H., Nicholson, A.C., Brunner, E. and Stansfeld, S.A. (1997) 'Low job control and risk of coronary heart disease in Whitehall II (prospective study)', *Br Med J*, 314: 558–63.

Colditz, G.A., Willett, W.C., Rotnitzky, A. and Manson, J.E. (1995) 'Weight gain as a risk factor for clinical diabetes mellitus in women', *Ann Intern Med*, 122: 481–6.

Department of Health (1988) *Public Health in England: The Report of the Committee of Inquiry into the Future Development of the Public Health Function*, London: HMSO.

Department of Health (1992) *The Health of the Nation*, London: HMSO.

Department of Health (1998a) *The Health of the Nation: A Policy Assessed*, London: Stationery Office.

Department of Health (1998b) *Inequalities in Health: A Report of the Independent Inquiry*, London: Stationery Office.

Department of Health (1999a) *Reducing Health Inequalities: An Action Report*, London: Stationery Office.

Department of Health (1999b) *Saving Lives: Our Healthier Nation*, London: Stationery Office.

Department of Health (2000a) *NHS Plan*, London: Department of Health.

Department of Health (2000b) *National Service Framework for Cardiovascular Disease*, London: Department of Health.

Department of Health (2001a) *Department of Health Select Committee on Health second report*, http://www.parliament.the-stationery-office.co.uk/pa/cm2001/cmselect/cmhealth/30/3007.htm.

Department of Health (2001b) *The report of the Chief Medical Officer's Project to Strengthen the Public Health Function*, London: Department of Health.

Gray, A. and Payne, P. (1993) 'The decline of infectious diseases: the case of England', in A. Gray and P. Payne (eds) *World Health and Disease*, Buckingham: Open University Press.

Lee, J.A. (2001) 'Myths, disasters and miracles', *Biologist*, 48: 52.

McKeown, T. (1979) *The Modern Rise of Population*, London: Edward Arnold.

Peters, J., Hedley, A.J., Wong, C.M., Lam, T.H., Ong, S.G., Liu, J. and Spiegelhalte, D.J. (1996) 'Effects of an ambient air pollution intervention and environmental tobacco smoke on children's respiratory health in Hong Kong', *Int J Epidemiology*, 25: 821–8.

Peters, J., Green, G. and Suokas, A. (2002) 'Evaluation of the initial impact of Sheffield Health Action Zone on reducing inequalities in health', Sheffield: CRESR.

Stokes, J., Noren, J.J. and Shindell, S. (1982) 'Definition of terms and concepts applicable to clinical preventive medicine', *J Community Health*, 8: 33–41.

Townsend, P., Phillimore, P. and Beatties, A. (1988*) Health and Deprivation Inequality*, London: Croom Helm.

von Mutius, E. (2001) 'Infection: friend or foe in the development of atopy and asthma? The epidemiological evidence', *Eur Respir J*, 18: 872–81.

WHO (2001) *Overview*. http://www.who.int/whr/.

Wilkinson, R.G. (1999) 'Putting the picture together: prosperity, redistribution, health and welfare', in M. Marmot and R.G. Wilkinson (eds) *Social Determinants of Health*, Oxford: Oxford University Press.

12 Human behaviour and the changing pattern of disease

Brian J. Ford

Introduction

The preceding chapters have offered us challenging insights into the development of the study of diseases and their impact on human health. This final chapter adopts a contemplative stance and considers the future. For fifty years, microbiology has dwindled to become a 'Cinderella' science. Young biologists have been discouraged from entering the discipline, since it has been tacitly assumed that the major infections were known and relatively little work remained to be accomplished. Virologists, when retired, found they were not replaced (Ford 2000a; Slade, personal communication 2001).

Much microbiology became subsumed into molecular biology, and the topic generally became regarded as unfashionable and *passé*. The levels of immunisation among the young fell to dangerously low totals, and when the BCG vaccine was proving difficult to source in Britain, wholesale vaccination of the teenage population was relegated to a matter of low priority. This is at variance with the realities of the subject, since the new millennium, arbitrary as the demarcation might be, may well herald an entirely new era of infection. We are encountering new diseases that may have implications for human health (such as BSE); recently identified pathogens (including *Helicobacter pylori* and *Campylobacter*); new types of well-known pathogens (for example, *E. coli* O157 H7); and a range of new means of transmission. From cling-film wrapped sandwiches, that offer a secure medium for cryophilic organisms like *Listeria*, to international transportation, which can bring people into contact with new diseases and can offer pathogens a unique channel through which to propagate an epidemic (in addition to global dissemination of the consequences), the new problems are largely undivined and widely ignored. This final chapter reviews a selection of the new problems, emphasising the grounds for the belief that a radically changed proactive stance to novel infection risks will be a necessary concomitant to a secure and civilised lifestyle in the future. Microorganisms are great opportunists, and the channels for transmission that are available in the modern world add greatly to the problems that will be encountered in the West.

Foodstuffs, traditional and novel

Disease has long been spread through food. Some of the earliest dietary strictures were based upon an ancient understanding of infections spread through the daily diet. The parasitic nematode *Trichinella spiralis*, spread by inadequately cooked pork, causes serious lesions and the fact has been known for thousands of years. It is to avoid such infections that the proscription of eating pig-meat amongst semitic and other races was first introduced. Infections with the beef tapeworm *Taeniarhynchus saginatis* are now rare. The primary host is mankind, while cattle are the secondary host. Current agricultural practices in the West make the closing of the cycle improbable, but plans to reuse human excrement as a soil conditioner and fertilising agent raise the possibility that infestations could reappear. The storage of chilled food allows some pathogenic organisms (like *Listeria*) to flourish; similarly, stored peanuts show a high incidence of growths of fungal organisms that are sources of aflatoxins. Although health and safety regulations for food workers have been widely advanced, the recorded incidence of food poisoning is steadily increasing throughout the Western world. We are all aware, from personal experience, that casual gastroenteritis is rarely reported to the public health authorities; thus the true extent of this problem is far greater than statistical evidence suggests. Much of the problem can be related to modern methods of food preparation and packaging, and it can be argued that insufficient emphasis is placed upon the health implications of newly introduced forms of food.

Notions of naturalness

There are some prevalent ideas on the supposed naturalness of foodstuffs that may predispose Western populations to harm. Nothing could be envisaged as more 'natural' than a hunk of bread and butter and a slab of cheese. In fact, these are the products of technology – traditional technologies, indeed, but profoundly unnatural processes none the less. The grinding of grains of the common bread wheat *Triticum aestivum* (= *T. vulgare*) to form a paste, which is then fermented through the action of the ascomycete yeast fungus *Saccharomyces cerevisiae* and baked in an oven, is far from a natural process. *T. aestivum* is itself a man-made species, unknown in nature; *Saccharomyces* is confined to the skin of fruit in its natural environment. Cheese is a product of a wholly unnatural set of processes; cow's milk is a natural food only for calves, and the fermented products known as cheeses are products of considerable technical refinement. Goat's milk is widely regarded as a more wholesome alternative, heedless of the fact that it is subject to fewer health controls and may be more frequently infected with pathogens. Health food fanatics are prominent consumers of synthesised meat substitutes such as Quorn, even though these are the products of extensive technological processing and have no counterpart in nature. Notions of 'naturalness' need objective reassessment.

Promiscuity

Since the 1960s the increased emphasis on sexual freedom has encouraged promiscuous behaviour. The Western world is characterised by a widespread acceptability of multiple sexual partners, and the fashionable nature of such attitudes means that the term 'promiscuity' is rarely encountered. This must be viewed in contrast to traditional societies, in which complex rituals to guard against promiscuity are widespread. The notion that the contraceptive pill enabled women to conduct relationships without fear of pregnancy circumvents the concern posed by a rising incidence of sexually transmitted disease. In consequence, a rising incidence of organisms such as *Chlamydia* is now apparent in young women. Female human anatomy makes it less likely that characteristic symptoms, such as a discharge, will be noticed. In consequence, we are witnessing a rise in the incidence of STDs. Although this applies to traditional diseases such as syphilis and gonorrhoea, it is in the newer conditions (including chlamydiasis and HIV) that the increase is most marked. The erosion of sub-Saharan societies by AIDS is particularly noteworthy; the very viability of many ancient communities is threatened by the widespread incidence of this tragic new disease.

Xenotransplantation

There is something so simplistic about the notion of xenotransplantation. A donor animal, customarily a pig, because of the anatomical similarity of its internal organs to those of humankind, is genetically modified so that surface markers on the cell surface will be recognised by the host (the human recipient) as self rather than non-self. The rearing of a pig as a source of a heart, kidneys, pancreas, etc., seems to offer hope for extensive organ transplantation without the complications of immunosuppression. In practice, it is not so simple, for mammalian species harbour viruses that may be symptomless and non-infective in the host, but potentially harmful to the recipient. A greater awareness of the extent of the risk of infectious disease in the course of this research might have prevented unduly optimistic forecasts of the progress of the research. A list of viruses known to be transmissible from pigs to humans in xenotransplantation is given in Table 12.1; the same authors list over 60 species of microorganisms that could be transmitted to humans via the same route. The occurrence of these potential pathogens does not inevitably lead to an inference that they will prove to be unavoidable hazards in practice; current research into the use of porcine stem cells in humans with nerve trauma is currently under way, and some indications of the safety of xenotransplantation will inevitably accrue from this line of investigation.

Potable water supply

There is a global shortage of potable water. In some arid nations, such as Saudi Arabia, reverse-osmosis desalination is used to produce potable water, but in other areas, extending from the Spanish coast to the western USA, the invasion of sea-

Table 12.1 Porcine viruses capable of transmission to humans following xenotransplantation

Porcine adenovirus
Porcine cytomegalovirus
Porcine rotavirus
Porcine endogenous and exogenous retroviruses
Aujeszky's disease virus
Japanese encephalitis virus
Encephalomyocarditis virus
Vesicular stomatitis virus
Swine vesicular disease virus
Foot-and-mouth disease virus
Rabies virus
Swine influenza virus
Swine parainfluenza-1 virus
[Malaysian Nipah virus, *vide infra*]
[Australian Paramyxovirus, *vide infra*]

Source: Borie *et al.* (1998), with permission from Slack Inc.

water into subterranean strata is already causing water supplies to contain unacceptably high levels of ionic sodium. The introduction of breast milk formula into tropical countries is posing problems of its own, for the water used to reconstitute the product is often contaminated with water-borne pathogens. Many areas are suffering a loss of forest cover, as wood is burned for fuel, and thus water is less often boiled. Water-borne diseases are posing an increased international problem. As we shall see, in Britain we have witnessed some outbreaks of *Cryptosporidium* infections and these may pose an important risk to the financial viability of the privatised water-supply industry.

Suprahygiene

A curious side-effect of the hygienic mode of living in the West has been a tendency for us to indulge in routine hand-washing, bathing and showering until we avoid exposure to so many antigens that we may challenge our immune systems too little. It can be argued that regular antigenic stimuli maximise the immune responses of the body, so that a highly sophisticated and ultra-clean community is at greater risk from opportunistic infection. There are several areas here that need to be addressed. One is that there are clearly communities of health-potentiating microorganisms that inhabit the skin, a lack of which might allow easier access to potential pathogens. Another is that early infections with microorganisms like *Cryptosporidium* traditionally produced a fleeting enteritis which conferred lifelong immunity. Keeping such challenges at bay may leave us increasingly liable to infection in the ultra-clean world of the future. Should this encourage us to reinvest in the old saying that a 'bit of dirt does no harm'? Perhaps not. Some of the current pathogens are well adapted to routes of transmission left open through poor hygiene, and in future we may need to maximise a systematic antimicrobial stance.

Internet marketing

Although some totalitarian regimes have tried to control the internet, and others (like China) attempt strictly to limit access to the world wide web, global access to the net is now the norm. The claim made for the internet is that it frees people to access global awareness. True as this is, it is less rarely recognised that it also offers opportunities for people to mail prohibited imports around the world and for customers to order potentially infected foodstuffs. Bush meats that are potentially contaminated with foot and mouth virus are widely exported in unlabelled packages. It may prove to be the case that the recent epidemic of foot and mouth in the UK originated in unlawful imports of infected meat from the Far East. There cannot be a single terrorist group anywhere in the world that has not watched the reports of the decimation of British agriculture with interest. Rather than expend large sums on missiles, extremists in future know that to close down a large portion of a national economy, all one need to do is to arrive with a virus-contaminated cloth in a pocket. The misuse of microbiological knowledge will open up unforeseen avenues of terrorism.

Commercial globalisation

Western economies traditionally functioned with a large reservoir of poorly paid workers, once on farms and in homes, as domestic servants, and more recently in factories and mines. The opening up of global markets means that the subservient classes exist as they did before, but are now geographically remote from those who benefit from their labours. The resulting global patterns of trade mean that potentially infected goods and products are sent worldwide as a matter of routine. The provisions of the UK health and safety legislation since the 1970s have bequeathed to us a framework that serves to reduce health risks from materials we purchase. Globalisation implies that products are imported from countries in which comparable safety regulations do not exist, and pose real hazards to public health. A recent outbreak of *Salmonella* in Australia was traced to imported halva (a paste made from sesame seeds).

Mass air transportation

A generation ago, air travel was still something of a luxury. The modern trend towards increasingly efficient and inexpensive air transportation encourages people from Western nations to travel to areas of the world where exotic diseases are endemic. At the same time, immunisation against infections is on the wane. The routine vaccination of schoolchildren using the BCG vaccine is not practised in the USA, and has been lacking in the UK for some decades due to 'problems of supply' with the vaccine itself. This leaves large populations at risk from tuberculosis. Worse still, the confinement of passengers within aircraft at altitude offers new opportunities for organisms to spread. Economies imposed by the air companies have led to a reduction in the rate at which fresh air is introduced into the cabin of an aircraft, which further increases the rate of contact between host and pathogen. Air travel is a major risk factor in epidemics due to novel pathogens (the SARS outbreak of April 2003 provided an example).

Reviewing the diseases

Many of the preceding chapters have examined specific areas of research and it would not be appropriate here to attempt a comprehensive review of the hazards we may face. A contemplative stance does lend itself to a representative review, and the following examples have been selected for special consideration.

Chlamydiasis

For many decades, genito-urinary infections with *Chlamydia pneumoniae* were hidden under the generic categorisation of NSU (non-specific urethritis). In 1985, I proposed that the term chlamydiasis be adopted to identify the widespread incidence of the organism. Between 1985 and 1990 the incidence of this condition increased five-fold in young American women. The incubation period is between two days and two months, and the symptoms may include a slight itching or burning which is often dismissed as 'mere cystitis'. A urethral discharge is frequently overlooked among normal vaginal discharges. The organism causes an inflammation of the fallopian tubes, and sterility is a common consequence of infection. The extensive pathology that results is promulgated by the vogue for sexual freedom. Although the infection is amenable to antibiotic therapy, infections may be relatively asymptomatic and thus damage may be considerable without the sufferer being aware of the existence of a potentially serious condition. Barrier contraception, which can militate against the transfer of pathogens as much as spermatozoa, needs to be seen as a crucial safeguard of personal health (Ford 1985).

Campylobacter

The genus *Campylobacter* has only been recognised in recent years. These are micro-aerophilic gram-negative organisms measuring 1.5–5 μm in length. Because of their unusual environmental preferences, the organisms did not appear in routine monitoring until the 1980s. Since studies began in the 1980s it has emerged as the leading cause of enteritis in the USA, and is known to be widespread throughout the Western world. Three species are of significance; *Campylobacter jejuni* is implicated in 90 per cent of outbreaks, while *C. coli* and *C. lari*, account for 9 of the remaining 10 per cent. The organisms are highly infectious; as few as 500 organisms may provide an inoculum. The disease features watery stools sometimes followed by bloody diarrhoea. The species listed above are non-invasive organisms, though *C. fetus* is normally an invasive bacterium. Some strains of *Campylobacter jejuni* are occasionally invasive. These strains cause pneumonia, meningitis, spontaneous abortion and severe forms of Guillain Barré syndrome. Three-quarters of the cases result from the consumption of food or water contaminated by animal waste, including fruit and vegetables, meat and poultry, shellfish and milk (Butzler 1984; Blaser *et al.* 1986; Nachamkin *et al.* 1992).

Listeria

Although the genus *Listeria* has been studied since the nineteenth century, it has recently attracted attention through its appearance in supermarket food products. *Listeria monocytogenes* is a gram-positive flagellated bacterium that has been isolated from a wide range of animal hosts and is found as a contaminant in soil. About 40 mammalian species have been shown to harbour *Listeria*, which can be isolated from intestinal samples of up to 10 per cent of the human population. *L. monocytogenes* has been contracted from unpasteurised (or ineffectively pasteurised) milk products such as soft cheeses, ice cream, raw vegetables, and all types of uncooked fish and meat, including uncooked sausages. An inoculum of less than 1000 cells is sufficient to induce infection. The symptoms are initially flu-like, progressing towards nausea, vomiting and diarrhoea. These typical forms of listeriosis can be followed by septicaemia and meningitis. *Listeria* provides a timely reminder of the exceptions to rules that we have to bear in mind. It is known that phagocytic cells protect the body from pathogenic microorganisms, and that refrigeration is a front-line defence against the proliferation of pathogens in foodstuffs. Neither is true in the case of *Listeria*. This organism can penetrate phagocytes and multiply within them. This remarkable property allows the organism access to the cerebrospinal fluid, and may also pass the placental barrier to produce an intrauterine infection of the foetus. Furthermore, the organism is able to multiply at temperatures around 3 °C, and it is thus able to proliferate in a refrigerator. Although it is the major outbreaks that attract attention (like the stillbirth epidemic of California in 1985, caused by soft Mexican-style cheese) there is a growing number of casual outbreaks. Currently there are around 2000 cases of *L. monocytogenes* reported in the US each year. Testing relies on culturing the organism and takes up to a week, though recombinant DNA techniques currently under development may reduce that to two days (Anon 2001).

Salmonella

This well-known organism has been thoroughly documented. It has caused political repercussions, as when Edwina Currie MP was obliged to resign after claiming in the House of Commons that 'most chicken production' in the UK was infected. If 'most' is taken to mean 'more than 50 per cent' then it may be that the statement was technically inaccurate. However, it is not entirely misleading, since the incidence of *Salmonella* infection is steadily increasing throughout the Western world. The genus includes several species of gram-negative rod-shaped organisms, all of which (except *S. gallinarum* and *S. pullorum*) are actively motile. *S. typhi* produces typhoid fever, but *S. paratyphi* can also produce a similar infection. As few as 15 cells may be sufficient to produce an infection with *S. typhi* and *S. paratyphi*. Although the initial symptoms include diarrhoea, cramping and headache, arthritis may supervene after a month. The sources of the bacteria include soil and water, factories and kitchen surfaces, and shellfish. In the retail environment, the organisms are particularly abundant in poultry and pork. *S. enteritidis* is particularly characteristic of poultry, and infected eggs are a hazard sufficient to have caused the general public health advice against the consumption of soft-boiled or poached eggs. It is now estimated that there may be half a million cases of salmonellosis in the UK each year, and up to 4 million in the USA. During the past decade, a six-fold increase has been recorded in the north-east United States. This is a potentially serious disease, and it is steadily on the increase.

Clostridium perfringens

This is a gram positive short rod-shaped non-motile sporing organism that is a common cause of food poisoning. As few as 10 organisms per gram of food can result in an infection with an incubation period of around ten hours. There are several strains of *Clostridium perfringens* (= *C. welchii* enteritis necroticans): type A1 produces enteritis with marked colic and diarrhoea, though diarrhoea and vomiting are usually absent. Type A2 produces an enterotoxin, while type C cause a necrotising enteritis. The type A bacteria can also cause wound infections that can culminate in gas gangrene or gangrenous cholecystitis. Currently, infections can be treated with penicillin, though some strains have been shown to have resistance to penicillin, tetracycline, erythromycin, chloramphenicol, metronidazole and clindamycin. The spores can survive for prolonged periods, and the bacteria have been shown to survive for up to a year in contaminated meat. Corned beef has sometimes been implicated in outbreaks. Recommendations include the storage of partly-used corned beef in a refrigerator, since the organism will not reproduce at or below 40 °F (4.4 °C) (Hatheway *et al.* 1980; Bean and Griffin 1990; Anon 1994a).

Bacillus cereus

This classical bacillus is widespread in nature. So abundant is the organism that it is an inevitable contaminant of food, though only toxin-producing strains cause illness. The symptoms of disease are similar to those caused by *Staphylococcus aureus* and *Clostridium perfringens*. Various toxins may be produced by the various strains of *B. cereus*; those of high molecular weight tend to produce diarrhoea, whilst those of lower molecular weight produce nausea and vomiting. The organisms produce toxins that reside in food, and thus trigger symptoms typically within 10 hours of consuming the contaminated food. Within 24 hours the symptoms typically subside. As is often the case in minor episodes of food poisoning, most cases are never reported to the authorities, so the accepted incidence (around 2 per cent of all outbreaks of food poisoning) is certainly an underestimate. Meat and fish have long been known to act as sources of the toxins, but in recent years rice has been more frequently incriminated (Anon 1994b).

Vibrio parahaemolyticus

In Japan, this is the most frequently encountered cause of casual gastro-enteritis. In the USA and Europe, the organism is common in estuarine waters and can be accumulated by filter-feeding molluscs. Most vibrios pass through the gastro-intestinal tract without causing problems for the host, but some organisms attach themselves to the intestinal wall and secrete a toxin. It is accepted that a dose of about 1 million organisms is needed to induce disease, though the number is greatly reduced in patients taking antacid preparations. The resulting disease manifests itself as diarrhoea and vomiting, abdominal cramps, headache and fever. In most cases the illness lasts for about three days and is mild to moderate, though some require hospitalisation. Related species can cause wound or ear infections, etc. (Anon 2001). Table 12.2 summarises food-borne *Vibrio* species.

Helicobacter pylori

The recognition of *Helicobacter pylori* in the human gastrointestinal tract was greeted with much interest, for it seemed to be the organism responsible for the great majority of duodenal ulcers. The discovery was made by Barry J. Marshall, an Australian microbiologist, who drank a broth culture of the organism and contracted gastritis as a result (Marshall 2001). It is said that a course of antibiotics resolved the illness, though colloquial reports suggest that the condition resolved spontaneously. Until *H. pylori* was recognised, the cause of ulcers was believed to be hyperacidosis consequent upon emotional stress or spicy foods, etc., acting upon the gastric and/or duodenal mucosa. However, a number of small spiral gram-positive bacteria were regularly observed in histological preparations of ulcerated gastric tissues, and these proved to be the *Helicobacter pylori* that has since been isolated in 90 per cent of sufferers from duodenal ulcers and 70 per cent of those suffering gastric ulceration. Aspirin and ibuprofen are claimed to be causally related to the remaining cases. There is also a close correlation with carcinoma, such that the WHO designated *H. pylori* as a 'category 1' carcinogen in 1994. The causative relationship between the bacteria and disease can be questioned, since the majority of people infected with the bacterium worldwide do not become ill, and in some surveys the levels of ulcer patients positive for the bacteria is much below 50 per cent. It can also be argued that the organism confers health by

Table 12.2 Food-borne *Vibrio* species potentially pathogenic to mankind

Vibrio alginolyticus
Vibrio carchariae
Vibrio cincinnatiensis
Vibrio damsela
Vibrio fluvialis
Vibrio furnissii
Vibrio hollisae
Vibrio metschnikovii
Vibrio mimicus

protecting against oesophageal carcinoma; in the areas of the West where the use of antibiotics has reduced the incidence of *H. pylori*, levels of cancer of the oesophagus are steadily increasing. It may be that the organism is a familiar inhabitant of the human gastrointestinal tract, where it can help protect against cancer, but which can become pathogenic if its environment is disturbed.

Certainly the effects of ulcers have been widespread. Among those who seem to have succumbed to the condition are Lorne Greene, Stonewall Jackson, Pope John Paul II, James Joyce, Ayatollah Khomeini and Imelda Marcos.

Cryptosporidium parvum

The type species of this genus is *C. muris* and it inhabits the gastric glands of several mammalian species (including laboratory rodents) but rarely infects humans. Although *C. muris* is frequently said to infect humans, the species involved is actually *Cryptosporidium parvum*. The genus is a member of the coccidia, and is a protzoan related to the gregarines. *Cryptosporidium* is a parasite of neonate mammals. Human infants traditionally contracted the infection through contact with farm animals, the transient diarrhoea that ensued conferring lifelong immunity. Interestingly, the human species is unusual in that we are able to contract the infection at any age. *C. parvum* enters the gut in the form of an oocyst containing four infective sporozoites that are liberated in the presence of digestive juices. The main site of infection is the ileum, where the parasites may be observed apparently attached to the lumenal surface of the mucosal cells. Electronmicroscopy reveals that the sporozoites are covered by cell membrane, and are actually intracellular parasites. Each subdivides to form eight merozoites which can colonise previously uninfected cells and thus perpetrate the infection. Cryptosporidia are not eliminated by chlorination of water, and large outbreaks have been reported from many parts of the Western world. It may be wondered why legal action for damages by business people has not been reported, but personal discussions suggest that the water companies resolve such disputes by out-of-court settlements, in which a non-disclosure clause is conditional (Clavel *et al.* 1996; Cordell *et al.* 1997; Atwill *et al.* 1999; de Graaf *et al.* 1999).

Verotoxin-producing Escherichia coli *(VTEC)*

Verotoxin-producing *Escherichia coli*, VTEC, is also known as *E. coli* O157:H7. This is a fast-emerging pathogen that was first identified in 1982. It appears that a strain of *E. coli* acquired genes coding for toxin production in the related coliform organism *Shigella*. From being a relatively obscure organism in the 1990s, *E. coli* O157:H7 has spread to infect the majority of cattle in the UK and USA, and is increasingly widespread around the world. The bacterium produces a bloody diarrhoea that can worsen and lead to widespread damage to internal organs. Farm animals may be asymptomatic carriers, and infants can also excrete the organism for some weeks after recovering spontaneously from infection. Older children (and adults) are much less likely to carry the organism and not show symptoms. Lavatory hygiene is important in preventing an outbreak, and so is the avoidance of raw or partly cooked beef. Although pasteurisation will kill *E. coli*, the organism has been spread through milk and fruit juice.

Antibiotic-resistant cocci

The gram-positive coccus *Staphylococcus aureus* is a common skin commensal, and was one of the first bacteria to be defeated by the newly introduced penicillin in the 1940s. Strains are now known that are resistant to conventional antibiotics and methicillin-resistant *Staphylococcus aureus* (MRSA, sometimes also known as 'multiple-resistant') are becoming common. Vancomycin, the antibiotic of last resort, is also now known to be ineffective in some cases, and vancomycin-resistant *Enterococcus faecalis* and *E. faecium* (VRE) are also becoming familiar. These organisms compromise the health of patients at risk, for example through a weakened immune system, and can produce life-threatening infections. Barrier nursing and personal hygiene are all important in the management of sufferers, and here we are reminded of the need for new antimicrobial agents. Many of these organisms can survive for up to a week on dry surfaces. They are now commonly found in the hospital environment, and iatrogenic infections are an increasing threat to our future.

Mycobacterium

Among those who have died of tuberculosis are:

> Frederick Chopin (d. 1849 aged 39) the composer
> Rene Laennec (d. 1826) discoverer of the acid-fast bacillus
> John Keats (d. 1821 aged 26) poet
> John Harvard (d. 1638 aged 31) founder of Harvard University
> Robert Louis Stevenson (d. 1894 aged 39) author
> Anton Chekhov (d. 1904 aged 43) playwright
> Max Lurie (d. 1966 aged 73) TB researcher
> Franz Kafka (d. 1924 aged 40) the surrealist author.

It was not unusual for entire families to be killed by TB. The Brontë family, descendants of the reverend Patrick Brunty (who changed his name in order to make it more socially acceptable), succumbed to TB: his wife Maria Branwell (d. 1821 aged 39) and offspring Maria (d. 1825 aged 12), Elizabeth (d. 1825 aged 11), Branwell (d. 1848 aged 31), Emily (d. 1848 aged 30) who wrote *Wuthering Heights*, Anne (d. 1849 aged 29) author of *Agnes Grey*, and Charlotte (d. 1855 aged 39) who wrote *Jane Eyre*.

Table 12.3 gives some traditional terms for TB.

The organisms that cause TB are acid-fast bacilli with a waxy coat that serves to make the disease difficult to treat. Immunisation using the BCG vaccine has been widely used in many countries to reduce the spread of the infection. On 20 November 1944 streptomycin was first used to treat tuberculosis, yet by 1947 the

Table 12.3 Traditional terms for tubercular infections

Consumption	Pulmonary tuberculosis
King's evil	Tuberculosis of cervical lymph glands
Long/Lung sickness	Pulmonary tuberculosis
Lupus vulgaris	Tubercular lesions of the skin
Mesenteric disease	Tuberculosis of abdominal lymph glands
Phthisis	Pulmonary tuberculosis
Pott's disease	Tubercular lesions of the spine
Scrofula	Tuberculosis of cervical lymph glands in young adults
White plague	Pulmonary tuberculosis
White swelling	Tuberculosis of the bone

first examples of resistance to streptomycin were being recorded. Since the introduction of streptomycin, many other antimicrobials have been brought into use. *Para*-aminosalycilic acid (PAS) was shown to have weak anti-tubercular effects in 1946, and from 1948 streptomycin was used with PAS to treat patients successfully. By 1951, streptomycin was replaced by isoniazid, and then came pyrazinamide (1954), cycloserine (1955), ethambutol (1962) and rifampicin (1963). Currently the aminoglycosides amikacin, capreomycin, viomycin and kanamycin are used with quinolones including ofloxacin and ciprofloxacin for resistant strains of the organism. The macrolides, which are also useful, await further clinical testing. In spite of these strides taken at the research level, strains of tubercle are now found that are resistant to all available agents. A generation ago the purpose of the Temple of Peace in Cardiff, Wales, was subject to reappraisal. The building had been provided by a charity to support the war against TB, and in the 1970s it was felt that there was no further use for the concept – TB was construed as having been beaten. In the global scenario, however, matters are unambiguously disturbing. One-third of the human population is infected with TB and one new individual is infected every second. An undetected victim will infect some 20 new people per year. Those infected with drug-resistant tubercle suffer a 70 per cent risk of mortality, and rates of infection in the Western world have been increasing since the mid-1980s. In the 1930s, levels of infection in the US were 175 per 100,000. By 1965 that had fallen to 2 per 100,000, but by 1992 the figure had risen to 52 per 100,000. The current invasion of resistant organisms is a cause for concern. In confined circumstances (such as an aeroplane, a subway or a conventional train) a large number of potential contacts could be involved, and in an era where the organisms are resistant to therapy the consequences could be serious (Nardell *et al.* 1986; Barnes *et al.* 1996).

Lyme disease

This debilitating chronic fever is caused by a tick-borne spirochaete named in 1984 *Borrelia burgdorferi*. The bacteria can be cultured on Barbour–Stoenner–Kelly agar but have a slow rate of growth, often requiring 24 hours for fission to be completed. *Borrelia burgdorferi* has been incriminated in Lyme disease in the USA, but *B. afzelii* and *B. garinii* are the causative agents of the disease in Europe. The syndromes produced by the species are somewhat at variance; arthritis seems to be more typical of the disease produced by *Borrelia burgdorferi*, and it also causes Lyme disease in Europe. In Asia, only *B. garinii* and *B. afzelii* cause Lyme disease in humans. Evidence is accumulating that these closely related, but different, spirochetes are associated with somewhat different disease expressions. Arthritis appears to occur more frequently following infection with *B. burgdorferi*, cutaneous symptoms are typical of *B. afzelii*, while neurological manifestations are more typical of *B. garinii* infections (Nadelman *et al.* 1990; Steere 2001).

Bartonella

Trench fever was the name given during the First World War to a louse-borne disease with symptoms including fever, rash splenomegaly and severe bone pain. It is caused by *Bartonella quintana*, and it is estimated that 1,000,000 troops were infected between 1914 and 1918. In the 1980s, the disease (which is also known as cat-fever) began to re-emerge, this time as an opportunistic infection in victims of AIDS in the USA, and later in France. Meanwhile the tick *Ixodes pacificus* from California has been shown to carry a reservoir of infection, and it may be that this accounts for episodic outbreaks. Research in Sweden on orienteers between 1979 and 1992 showed that 16 suffered heart attacks in a sample where one might be anticipated. *Bartonella* was incriminated in these cases, and the organism is already known to cause cardiac damage in cats. This disease may be more widely implicated in public health than is currently realised, and appears to be increasing in its geographical spread. Antibiotic therapy with azithromycin, doxycycline, erythromycin or tetracycline is claimed to be effective (Relman 1995; Stein and Raoult 1995; Jackson *et al.* 1996).

Emergent viruses

Human communities are being colonised by organisms that have not been encountered before. *Campylobacter*, VTEC (*E. coli* O157:H7), MRSA, and multiple drug-resistant *Mycobacterium* are among the bacterial pathogens that have emerged within the last generation. Other current threats to human health are of more ancient lineage. *Bacillus anthracis*, recently used in mail-borne terrorism, is not infectious, person to person, but many of these recently emerged organisms have the capacity to trigger epidemic outbreaks. It is when virus outbreaks are considered, however, that an impression may be gained of the magnitude of the potential problems in a new era. A representative selection of these may be contemplated, including, as a finale, some viruses that have been identified within the last few years.

Hantavirus

Korean haemorrhagic fever, caused by Hantavirus, was first recognised by Western scientists during the Korean war. The disease is spread by rodents. In recent years, cases have been reported from Argentina and Chile, and now there have been reports from the USA, notably California. The strains of the virus cause different syndromes, including Hantavirus Pulmonary Syndrome (HPS) which has been found in the USA, and Haemorrhagic Fever with Renal Syndrome (HFRS) in which renal failure, haemorrhage, and shock occur in sequence. Levels of mortality are about 80 per cent in the USA. In the last fifteen years there have been some 200 cases in Argentina, less than 30 in Chile, and a handful in Brazil and in California. The incubation period can be over a month, followed by the sudden onset of severe respiratory distress and likely death. On 23 May 2001 the Centers for Disease Control and Prevention (CDC) announced that they had begun to receive reports from authorities concerned over the fate of a stock clerk who became infected with Hantavirus while working in a storeroom. According to the e-mail message, the infection resulted from exposure to dried rodent droppings that were contaminated with Hantavirus. It was a hoax. As seen in more recent times, biological agents are peculiarly susceptible to misappropriation by hoaxers. A greater familiarity with the behaviour and nature of potential pathogens on the part of the public is the surest way of keeping such matters in perspective.

West Nile virus

This is an emergent disease with 10 per cent mortality that is causing much interest in the USA since it has begun to cause infection on the East Coast in the last few years. In 1999 there were 58 cases and seven deaths, with much attendant publicity in the press and broadcast media across the Western world. This may be contrasted with the position in 1997, when there were over 500 cases in Romania and little attention was paid by America or Britain. The disease emerged in New York City on 4 August 1999, and on 31 August an 80-year-old man died of encephalitis. This was the first time the virus had been detected in the Western Hemisphere. The secondary host for the virus is avian, and about 20 species – from the crow to the bald eagle – have been shown to harbour the virus. Mosquitos transmit the virus to the human host. The effect of global warming would encourage the proliferation and geographic spread of the insect vector, which allows us to infer that such outbreaks may become more prominent in future decades (Anon 1999b; Hubalek and Halouzka 1999).

Lassa fever

This disease was first described in the 1950s, though the virus, a member of the Arenavirus group, was not recognised until 1969. Human carriers may be entirely asymptomatic, but in others the disease exerts a powerful and devastating effect. A gradual onset of headache, nausea, cough, vomiting and diarrhoea progresses to shock, effusion and haemorrhage with encephlopathy. Fifteen per cent of hospitalised patients die, and those that survive may suffer deafness, alopecia and a loss of manual coordination. Ribavirin, an antiviral, may prove to be effective in therapy if administered within the first six days of onset. The disease is found in

Guinea, Liberia, Nigeria and Sierra Leone. The secondary host, recognised during the 1970s, is the multimammate rat *Mastomys natelensis*. There is no vector, the virus being acquired by inhalation of dried excreta, etc., from the rodents. Once established in the human host during the acute phase the disease is highly contagious, being spread easily to unprotected hospital personnel, for instance. The incubation period is one to three weeks. There was a major outbreak in Sierra Leone in 1996–97. A total 823 cases were reported with 153 deaths (18.6 per cent mortality).

Ebola and Marburg haemorrhagic fevers

The existence of the highly dangerous Filoviridae emerged in 1967 when scientific staff in Belgrade, Yugoslavia and Marburg, Germany, became infected with a virus contracted from tissue samples from the African green monkey *Circopithecus aethiops*. Of the 25 primary cases, seven died. Sporadic outbreaks have since occurred in Zimbabwe, Kenya and South Africa. Subsequently, the related Ebola virus gave rise to outbreaks in Zaire and Sudan in 1976. Over 500 cases were reported, with a mortality rate of 53 per cent in Sudan and 88 per cent in Zaire. In November 2000 an outbreak in Uganda of 280 cases of Ebola haemorrhagic fever resulted in 89 deaths, a mortality rate of 24.9 per cent. No association between Ebola and monkeys is known. Further discoveries of Filoviridae have been made in the cynomolgus monkey *Macaca fascicularis* that is native to the Philippines. These are the Reston viruses which, though serologically similar to Marburg and apparently transmissible to humans, do not cause a disease in the human host. The symptoms and course of the diseases are legendarily unpleasant. After an incubation period of 4–16 days fever, headache and myalgia become manifest, soon followed by nausea, dehydration and diarrhoea. Haemorrhage appears from the orifices, lungs and into the abdomen from the gastro-intestinal tract. It has been shown that Novalgin (Metamizol, Antipyrin) can reduce intravascular coagulation in Marburg disease. However, vaccination of experimental animals with antigen and with inactivated whole virus particles has shown no prophylactic benefit following a challenge with live virus. A vivid portrayal of an epidemic in the California hills in the Hollywood film *Outbreak* was drawn to the conclusion that a nuclear attack on the district was the only way to curtail it. Clearly, in the minds of the public, such viruses pose serious threats.

Dengue fever

This is the most widespread of all haemorrhagic fevers across the world. Fortunately, most cases do not progress to this stage, but unfortunately this is the most rapidly spreading insect-borne disease known to us. The condition was first described by Western medical science in 1780, though modern reports of epidemics date only from 1949. Over two billion people live in at-risk areas, and tens of millions of new cases occur each year. Haemorrhagic fever effects annually perhaps a quarter of a million people. No vaccine is available, though an experimental attenuated candidate vaccine has been investigated in Thailand. There are four serotypes of Dengue fever, and an infection with one serotype, far from conferring resistance, seems to increase the likelihood of a haemorrhagic fever in any subsequent infections. The disease is widespread across large areas of Africa, Asia and South America (including the Caribbean). The vector is the mosquito *Aëdes aegypti*, and the use of a reliable insect repellant is the most effective prophylatic measure travellers can take. Dengue fever is characterised by a sudden high fever, myalgia, nausea and retro-orbital pain, often accompanied by bradycardia. In most cases, symptoms persist for a week (hence the common name, 'seven-day fever'). In some cases the immune system collapses, with a further increase in temperature, shock convulsions and a fatal outcome. Haemorrhagic dengue manifests itself as prostration and bleeding from the gastro-intestinal tract and elsewhere, and has 5 per cent mortality. In recent years the disease has been appearing in states in which it has previously been undetected, including Venezuela (1990), Brazil (1991), Djibouti (1992), and Pakistan, Saudi Arabia and Nicaragua (1994).

Nipah virus

In March 1999, an outbreak of a new virus from pigs was reported in Malaysia (and Singapore). The disease manifested itself as an acute encephalitis and respiratory illness that often resulted in death. The cause proved to be a hitherto unrecognised paramyxovirus that has been named Nipah virus. Once the connection with pigs was recognised, abattoirs were closed in Singapore, and no further cases were reported. In Malaysia, by 27 April 1999 there had been 257 cases of febrile encephalitis recorded by the Malaysian Ministry of Health (MOH), 100 of which had a fatal outcome. Some further cases were reported in Negeri Sembilan and Selangor. The most likely source of the virus is pigs; in the Negeri Sembilan outbreak, almost 90 per cent of the victims reported they had recently been close to pigs; two-thirds of the patients reported that the pigs with which they had been in contact had appeared unwell. The virus has not been shown to be transmissible from person to person, and it was controlled by the slaughter of 900,000 pigs in the affected areas (Anon 1999a).

Paramyxovirus

In 1994, a hitherto unknown equine morbillivirus (EMV) caused an outbreak of a severe disease in horses and humans. Levels of mortality appear to be high: of 21 horses found to be infected at Brisbane, 14 died (or were put down) due to severe respiratory distress; of the two human contacts suffering a similar disease, one died. The source of the virus appears to be the fruit bat or flying fox *Pteropus*. Two outbreaks were reported within a month of each other: one at Brisbane, the other at Mackay; towns separated by 1000 km. In the Mackay incident two horses became ill and both died. One person also died of a relapsing encephalitis. The first case in both outbreaks appears to have been a mare late in pregnancy and grazing on open pasture, but no connection has been shown to exist between the two outbreaks.

Serological examination of samples from 46 wildlife species showed that approximately 10 per cent of *Pteropus* carries antibody to the virus (Murray *et al.* 1995; Selvey *et al.* 1995).

Other causative agents

A range of new, re-emergent or emergent infections that are of growing importance in the field of public health have been examined. The examples discussed above are viral or bacterial diseases. There are others of interest in which causative agents are either non-genetic or still unknown. The general inference that must be drawn is that infectious disease, far from being a field in which most of the research has been accomplished, is one in which new challenges are arising. The examples that follow will exemplify some current problems.

vCJD (variant Creutzfeldt–Jakob disease)

Following the recognition of bovine spongiform encephalopathy (BSE) in 1986, a novel spongiform encephalopathy was recognised in human patients in 1996. Although in symptomatic terms it has more in common with kuru than with classical Creutzfeldt-Jakob disease, it was dubbed nvCJD (new variant Creutzfeldt-Jakob disease), a term recently shorted to vCJD (variant CJD). The origin of both diseases remains problematical. It has been argued that a change in the rendering processes for meat and bone meal led to an escape into animal feed of the causative agent, a modified prion. The diagnosis of cases of human spongiform encephalopathy a few years after surveillance began has led to the general acceptance that BSE in infected cattle led to vCJD in humans, although clearly no precise causal link has been demonstrated. The saga of vCJD posed many curious questions, and showed how official sources (from ministry to government) conspired to launder realities for short-term commercial expediency. Although much was said of the government's determination not to make similar mistakes in future, the outbreak of foot and mouth (= hoof and mouth) in February 2001 showed a similar inability to grasp the realities of disease control. In October of the same year it announced that the BSE-positive samples of sheep brain tissue were erroneously ascribed – bovine samples had mistakenly been analysed instead. It is examples of this sort that encourage me to feel that present-day governments are ill-equipped to handle disease outbreaks. A fundamental reappraisal of criteria in cases like these is a matter of utmost urgency. Future outbreaks could have more far-reaching consequences (Tyrrell 1994; Ford 1996).

Kawasaki syndrome

Academics and clinicians in public health will recall the claim that rheumatic fever is the leading cause of acquired cardiac disease in children across the Western world. This is no longer the case; Kawasaki disease, which was first described in Japan in 1967, has now taken over the title. Eighty per cent of the cases occur in children aged under five years, and it is characterised by pyrexia, reddening of the eyes, lips, skin and extremities, and cervical lymphomegaly. After a week or more, the skin of the peripheries may peel and arthritic symptoms appear. Although precautions against Reyes syndrome must be taken (e.g. offering influenza and chickenpox vaccine), therapy with aspirin is considered advantageous. Gamma-globulin, given in large doses within the first ten days of the disease, is believed to be important in aiding recovery. Crucially, a weakening of the coronary arteries is a consequence in some 20 per cent of sufferers. This can cause sudden collapse and death in later childhood. The mortality rate is 0.5 per cent. What might be the cause? The erythroderma and peeling of the skin suggest some form of toxic shock syndrome, and research has been directed to eliciting signs of an infectious viral or bacterial pathogen. However, Kawasaki disease is a condition of growing importance and with a potentially tragic outcome, yet it still lacks an identifiable cause (Shulman *et al.* 1995).

Priorities for the future

Human society has developed through the conquering, or at least the management, of disease. With occasional perturbations, like the bringing of measles to the Eskimo peoples, or the importation of syphilis into Europe, the graph of infectious disease has had a downward path, so that each generation has tended to have less to fear from infectious disease. Scourges of previous decades, from scarlet fever and polio-myelitis to mumps and measles, have faded within living memory. At the beginning of a new millennium, we are, for the first time, facing a significant change in that downward graph; now we are facing new threats, as hitherto unknown diseases emerge to threaten our complacency. In many cases this is due to human intervention, for microorganisms are quick to exploit the loopholes that our innovations (like supermarket food, or mass international travel) can offer. At a deeper level, this problem is exacerbated by a sense that the study of bacteriology was somehow unfashionable and *passé*, coupled with a dearth of understanding on the part of politicians and the public alike. Departments of microbiology are poorly funded, and public health services are being curtailed. As this chapter has shown, we are facing a series of novel problems that only a fundamental change of attitudes, and a reappraisal of our criteria, can ameliorate. There is a tendency to regard warnings of impending problems as scare-mongering, but such a criticism cannot apply to cases like these. These are diseases which we are facing now, and in greater amounts than we once were. Millions of people are ill; urgent action is needed to prevent that number from dramatically increasing. Human behaviour must change, and change soon.

The spread of vectors

Global warming may have an effect on the spread of vectors. The anopheline mosquito is spreading northwards toward northern Europe, and other insect species are spreading further. The Asian tiger mosquito *Aëdes albopictus* is becoming widespread in the USA, although it is not a native species. *A. albopictus* first appeared in Texas in 1982. Larvae were present in pools of water lying in used tyres that were being imported from the East. The mosquito has since spread to about 20 states in the USA. Among the diseases that it is capable of transmitting are dengue fever, yellow fever, Mayaro, Venezuelan and eastern equine encephalomyelitis. It is noteworthy that *A. albopictus* is somewhat hardier than *A. aegypti*, and may therefore extend the range of insect-borne infections. Until the over-use of DDT led to its prohibition, extensive programmes of vector eradication had done much to reduce their extent, and vector distribution in the present day is far higher than it was. A notion that was prevalent in the USA in the post-Second World War period was of a nation freed from all insects, which were seen at the time as nothing more than superfluous. Ill-informed as this might be, we do need to investigate means of controlling the insect transmission of human disease, and can anticipate that research can throw light on the most suitable means of limiting the spread of such potentially hazardous insect species without compromising the crucial importance of the insect world in the management of the global ecology.

Expert misinformation

The public are ill-equipped to comprehend microbiological matters. Little of relevance is taught at school, and many of the popular reference sources are inaccurate or misleading. At the Royal Society of London, a document for the public that set out to explain the background to spongiform encephalopathies stated that: 'The human form [of spongiform encephalopathy] is CJD', heedless of kuru, fatal familial insomnia, and Gerstmann–Sträußler–Scheinker syndrome. A recent account by the medical columnist of a major Sunday newspaper in Britain described the agent of tuberculosis as 'a virus'. Meanwhile, a student textbook describes how viruses become 'accustomed to antibiotics'. When the authorities write with such a disregard for elementary realities it is unreasonable to expect the public, or policy-makers, to have a detailed understanding of the issues involved (Ford 1975; Reynoldson 1996; Henderson 2001).

Education – sexual and otherwise

From the earliest age, young people should be given a full understanding of the ubiquitous nature of microorganisms, their vital role in the cycles of the environment, and the sheer sense of wonder that watching life under a microscope can convey. A topic like 'public health' should feature in the syllabus along with current topics like 'environmental studies' and 'media and communication'. Teaching medicine to medical undergraduates is not enough; the subject should appear on the syllabus from the age of eleven. Hygiene should be a matter of course. Sexual education needs to pay far greater emphasis on the transmission of pathogenic microorganisms through promiscuity. Although the term is rarely encountered, being regarded as somehow judgemental and unfashionable, it remains a fact that the social acceptability of multiple sexual partners in short-term and uncommitted relationships provides many opportunities for the dissemination of pathogens. The decimation of many African nations, and the tragic loss of so many young lives in the Western world and elsewhere, is testimony to the effectiveness of sexual transmission for HIV, and the bacterial diseases – syphilis, gonorrhoea – are proliferating once more as chlamydiasis is taking a hold on young lives. An acceptable notion of casual sex has been inculcated in the young, with no concomitant awareness of the hazards to health that this implies. The availability of the contraceptive pill was heralded as the key to an unfettered sex-life for women. The development was hailed as a timely breakthrough for the drug companies, and a time for excitement in young men. To the microorganisms that rely upon sexual transmission, it was the most propitious development of all.

Immunisation

The widespread application of vaccination, promulgated by Edward Jenner, led to the eradication of smallpox. Subsequently, stocks of the virus became confined to three nations: Russia, the USA and South Africa. South Africa was induced to give up her stocks in 1983, leaving the two super-powers as holders of the virus in high-security establishments. On 30 June 1999, Russia and the USA claimed to have destroyed their remaining stocks. Research into the uses of smallpox virus as an agent of warfare was conducted by many nations in the 1960s, though in 1972 the Biological and Toxic Weapons Convention was signed and such research was halted. However, it was claimed that Russia had equipped warheads with smallpox virus, and the nations that have been claimed to hold clandestine stocks of the virus include North Korea, China, Israel, India, Libya, Syria, Iraq and Iran. The evidence for such claims is slight, though the existence of related pox viruses that are amenable to genetic modification is sufficient for us to remain aware to the possibilities of bioterrorism using such agents. In the case of smallpox, the abandonment of routine vaccination was clearly justified at the time, although vaccine may be needed should an unauthorised source of the virus manifest itself and be released.

The failure to continue BCG vaccination in the UK, by contrast, has less scientific support, particularly in an era when antibiotic-resistant strains of tubercle are becoming widespread. If resistant strains become increasingly widespread in the USA, routine vaccination may need to be offered to Americans. In other cases, lowered rates of protection are elective.

Concerns felt by parents over the safety of the triple MMR vaccine have led to greatly reduced levels of protection. Although unequivocal evidence for the relationship between MMR and subsequent brain damage is not available, it could be argued that the immune system is not ordinarily simultaneously challenged by disparate antigens, and an option for the administration of the vaccines in spaced doses should be available for parents who desire this. The British authorities have proscribed this method of administration, a form of heavy-handedness that is leading to dangerously low levels of protection. The recalcitrance of officialdom is leading children to remain unprotected when separate vaccinations would be acceptable to many of the families who reject the triple vaccine. In the present situation, the public need to be vaccinated with every available form of treatment for each prevalent disease. It is incumbent upon medical authorities to take every measure possible to maximise compliance.

Public preoccupations

Today's public are concerned about safety. They read food labels, often uncomprehendingly. They are perturbed by what are construed as 'scientific' ingredients. The E-number system, widely used in Europe during the 1980s, has been abandoned. The listing of ingredients by number intimidated the public, and people used to set out to avoid food 'full of E-numbers', heedless of the fact that wholly beneficial ingredients (vitamins, for example) were among those on the list. The current vogue is for 'organic' foods, however construed. Yet labelling of food products is often suspect. Although it is widely assumed that the label tells the truth, there are companies in existence who replace labels, who re-label out-of-date food, and who falsify descriptions. It would be helpful if reporters began to investigate some of these. A recent case in Britain that involved the public resale of chicken meat condemned as unfit for human consumption has been widely reported; such cases are unlikely to be the rarity that this isolated case implied. We could seek to involve the public far more in the realities of microbiology. Such topics should be taught at school, for they would establish codes of behaviour of immeasurable benefit in adult life. The use of dilute hypochlorite as a kitchen sterilant, would be valuable if it could be well understood, and elsewhere I have proposed that a sterilising bowl may become a feature of the kitchen facility in homes of the future (Ford 2000b).

It needs to be understood that the use of washing machines that run at blood heat, less than 40 °C, does not destroy pathogens. Although the surfactant properties of detergents may well act against pathogens, washing machines at such low temperatures are afflicted by a build-up of greasy waste. A high temperature wash may prove to be a useful safety precaution. Airline operators may have to reconsider the rate of atmosphere exchange within their aircraft if it is to be ensured that the body of the plane does not become a re-infection chamber for healthy passengers travelling with an infectious individual (particularly those with a resistant strain of tubercle, for instance). Governments need to adopt a more proactive stance, and need to be closer to science and to scientists if they are to make sense of future problems. Following the debacle of bovine spongiform encephalopathy in the UK, much was heard of governmental intentions to handle matters more effectively in the future, though their subsequent management of foot and mouth has been widely criticised for a failure to apply the lessons learned. In an era when microbiological hazards are multiplying, the behaviour of governments that are ignorant of the principles of biology can no longer be relied on.

References

Anon (1994a) *Health Canada*, 43: 137–8, 143–4.

Anon (1994b) *Morbidity and Mortality Weekly Report*, 43: 18, Centers for Disease Control and Prevention.

Anon (1999a) 'Outbreak of Hendra-like virus, Malaysia and Singapore 1998–1999', *MMWR*, 48: 265–9.

Anon (1999b) *Morbidity and Mortality Weekly Report*, 48 (39): 890–2, Center for Disease Control and Prevention, Atlanta, GA.

Anon (2001) *Food-borne Pathogenic Microorganisms and Natural Toxins Handbook*, U.S. Food and Drug Administration Center for Food Safety and Applied Nutrition.

Atwill, E. R. *et al.* (1999) 'Age, geographic, and temporal distribution of fecal shedding of Cryptosporidium parvum oocysts in cow–calf herds', *American Journal of Veterinary Research*, 60: 420–5.

Barnes, P.F., El-Hajj, H., Preston-Martin, S. *et al.* (1996) 'Transmission of tuberculosis among the urban homeless', *Journal of the American Medical Association*, 275: 305–7.

Bean, N.H. and Griffin, P.M. (1990) 'Foodborne disease outbreaks in the United States, 1973–1987: pathogens, vehicles, and trends', *Journal of Food Protection*, 53: 804–17.

Blaser, M.J., Perez, G.P., Smith, P.F., Patton, C.M., Tenover, F.C., Lastovica, A.J. and Wang, W.G. (1986) 'Extra intestinal *Campylobacter jejuni* and *Campylobacter coli* infections: host factors and strain characteristics', *Journal of Infectious Disease*, 153: 552–9.

Borie, D.C. *et al.* (1998) 'Microbiological hazards related to xenotransplantation of porcine organs into man', *Infection Control and Hospital Epidemiology* 19 (5): 359.

Butzler, J.P. (1984) *Campylobacter infection in man and animals*, Boca Raton, FL: CRC Press.

Clavel, A. *et al.* (1996) 'Seasonality of cryptosporidiosis in children', *European Journal of Clinical Microbiology of Infectious Diseases*, 15: 77–9.

Cordell, R.L. *et al.* (1997) 'Impact of a massive waterborne cryptosporidiosis outbreak on child care facilities in metropolitan Milwaukee, Wisconsin', *Pediatric Infectious Disease J*, 16: 639–44.

de Graaf, D.C. *et al.* (1999) 'A review of the importance of cryptosporidiosis in farm animals', *International Journal of Parasitology*, 29: 1269–87.

Ford, B.J. (1975) 'Microscopic blind spots', *Nature* 258: 469.

Ford, B.J. (1985) 'Sexually transmitted disease', in J. Bevan (ed.) *Sex and Health*, London: Mitchell Beazley, pp. 84–137.

Ford, B.J. (1996) *BSE – the Facts*, London: Transworld.

Ford, B.J. (2000a) (ed.) *The First Fifty Years*, London: Institute of Biology.

Ford, B.J. (2000b) *Future of Food*, London: Thames & Hudson.

Hatheway, C.L., Whaley, D.N. and Dowell, V.R. Jr. (1980) 'Epidemiological aspects of *Clostridium perfringens* foodborne illness', *Food Technology*, 34: 77–9.

Henderson, R. (2001) 'Tuberculosis – the case of the kindergarten teacher', *Sunday Times Magazine*, 27 May.

Hubalek, Z. and Halouzka, J. (1999) 'West Nile fever, a re-emerging mosquito-borne viral disease in Europe', *Emerging Infectious Diseases*, 5 (5), Center for Disease Control and Prevention, Atlanta, GA.

Jackson, L.A., Spach, D.H., Kippen, D.A., Sugg, N.K., Regnery, R.L., Sayers, M.H. and Stamm, W.E. (1996) 'Seroprevalence of *Bartonella quintana* among patients at a community clinic in downtown Seattle', *Journal of Infectious Diseases*, 173: 1023–6.

Marshall, B.J. (2001) website URL: http://www.vianet.net.au/~bjmrshll/.

Murray, P., Selleck, P., Hooper, P., Hyatt, A., Gould, A., Gleeson, L. *et al.* (1995) 'A morbillivirus that caused fatal disease in horses and humans', *Science*, 268: 94–7.

Nachamkin, I., Blaser, M.J. and Tompkins, L.S. (1992) (eds) Campylobacter jejuni *Current Status and Future Trends*, Washington, DC: American Society for Microbiology.

Nadelman, R.B., Pavia, C.S., Magnarelli, L.A. and Wormser, G.P. (1990) 'Isolation of *Borrelia burgdorferi* from the blood of seven patients with Lyme disease', *American Journal of Medicine*, 88: 21–6.

Nardell, E., McInnis, B., Thomas, B. and Weidhaus, S. (1986) 'Exogenous reinfection with tuberculosis in a shelter for the homeless', *New England Journal of Medicine*, 315: 1570–5.

Relman, D.A. (1995) 'Has trench fever returned?', *New England Journal of Medicine*, 332: 463–4.

Reynoldson, F. (1996) *Medicine Through Time*, London: Heinemann.

Selvey, L., Wells, R.M., McCormack, J.G., Ansford, A.J., Murray, P.K., Rogers, R.J. *et al.* (1995) 'Outbreak of severe respiratory disease in humans and horses due to a previously unrecognised paramyxovirus', *Medical Journal of Australia*, 162: 642–5.

Shulman, S.T. *et al.* (1995) 'Kawasaki disease', *Pediatric Clinics of North America*, Oct: 1205–22.

Steere, A.C. (2001) 'Lyme disease', *New England Journal of Medicine*, 345: 115–25.

Stein, A. and Raoult, D. (1995) 'Return of trench fever', *Lancet*, 345: 450–1.

Tyrrell, D. (ed.) (1994) *Transmissible Spongiform Encephalopathies, A Summary of Present Knowledge and Research*, London: HMSO.

Index